Onzième Congrès International de Stratigraphie et de Géologie du Carbonifère

Editeé par Jin Yugan & Li Chun

Beijing

August 31 — September 1, 1987

Compte Rendu

Tome 1

Nanjing University Press

1991

Responsible Editors Deng Longhua
Xu Shanhong

Published by Nanjing University Press
Printed by Amity Printing Co. Ltd., Nanjing

ISBN 7-305-00510-X/P·34

CONTENTS

PLENARY LECTURES

Openning ceremony at the People's Great Hall

Field excursion in Xinjiang

Highlights of reception

Chinese hosters at reception

XIe Congrès International de Stratigraphie et de Géologie du Carbonifère Beijing 1987,
Compte Rendu 1 (1991): 1-7

A Briefing of the XIth International Congress of Carboniferous Stratigraphy and Geology

Jin Yugan and Li Chun

The International Congress of Carboniferous Stratigraphy and Geology initiated in Heerlen, the Netherlands, in 1927 is one of a few international gathering of geologists with a long history. It has played an important role as a motivation in promoting the search for fossil fuels and mineral resources related to the Carboniferous as well as the basic research of Carboniferous geology. Previously, the Congresses were exclusively held in the countries of western Europe, USA and USSR, though its scope had been globalized according to the geographical areas covered by the papers presented and the nations of delegations since the fifth Congress in Paris. Having noted that the participants of the Congress intended to visit the Carboniferous in countries of Asia and the southern hemisphere, Dr. Garcia-Loygorri, Chairman of the Spanish Organizing Committee for the Xth Congress, wrote to the President of the Chinese Academy of Sciences and the Secretary General of the Geological Society of China respectively in July. and November, 1982 to explore the possibility of hosting the XIth Congress in China. Thanks to implementation of the reform and open policy, the Chinese government has been encouraging scientific exchange between Chinese and foreign experts since 1979. Moreover, the Carboniferous strata are well developed and widespread in China, which provide the main basis for Chinese coal industry and other important mineral resources. A considerable amount of work on stratigraphy, geology and coal resources has been done in the last three decades. Therefore, a favorable response was soon given to our Spanish colleague. During the Xth Carboniferous Congress in Madrid, Prof. Yang Jingzhi was invited to fill the seat of Chinese representative in the International Permanent Committee for the Carboniferous Congress. He presented an official requisition of China for hosting the XIth Carboniferous Congress in Beijing in 1987, which was unanimously approved by the Committee. This Congress held in China marks the first time that an Asian country outside of Europe and North America become its host. To Chinese geologists, this Congress, the first large international conference on geological sciences ever held in China, implies rapidly broadening communication channels with foreign colleagues.

Planning of the XIth-ICC

The Chinese Organizing Committee with Prof. Chen Yuqi, the President of Chinese Geological Society, and Prof. Liu Dongshen, the Head of Chinese Association of Sciences and Technology as Chairman and Secretary General, respectively, and the Chinese Program Committee with Chairman Prof. Yang Jingzhi, Chairman of Committee on Carboniferous Stratigraphy, All-China Commission on Stratigraphy (ACCS), were set up at the beginning of 1985. Both committees contain representatives from the Palaeontological Society of China, the Geological Society of China, the Chinese Society of Coal, and the Chinese Society of Petroleum. The Scientific Secretariat based at the Nanjing Institute of Geology and Palaeontology, Chinese Academy of Sciences, is responsible for carrying out the plan formulated by both committees. It was firmly supported by an active group of experts of Carboniferous-Permian stratigraphy and palaeontology, including Profs. Yang Jingzhi, Li Xingxue, Chairman of Committee on Permian Stratigraphy, ACCS, Sheng Jinzhang, Chairman of International Subcommission on Permian Stratigraphy, IUGS, and Wu Wangshi, Chairwoman of International Working Group on Carboniferous/Permian Boundary. Both committees met three times and finally declared at a meeting held two days before the opening of the Congress that they have successfully finished their task and should shift to an executive committee to conduct the Congress. Prof. Wu Heng, the Chairman of the All-China Commission on Stratigraphy was officially invited as the President of the XIth Congress.

The organizing work attracted attention from numerous Chinese geologists, since they were eager to take this unique opportunity to share new scientific data and theories with foreign colleagues, which will be a help in better understanding the Carboniferous geology of China. The Carboniferous in China is characterized by a dominance of Tethyan deposits with some others closely related to the Gondwana in Xizang and to the Boreal along the northern border region. Fully developed marine sequences enable them to offer candidate sections for the stratotype for the base, top, and internal boundaries of the system. Special national meetings had been held in order to collect papers and publications to be presented during the Congress. A stratigraphical scheme of the Carboniferous in China was formulated at a meeting sponsored by the All-China Commission, and the Chinese Program Committee for use in those publications. The Chinese Society of Coal and the Geological Society of China held a meeting on the Carboniferous-Permian coal-bearing strata in China to publish a volume of selected papers. A highlight of this Congress was the presentation of publications by Chinese geologists on the Carboniferous stratigraphy, geology and coal resources of China. Detailed reviews of Carboniferous stratigraphy in China were presented to participants in the form of three books from the Nanjing Institute of Geology and Palaeontology: 1) *Carboniferous Stratigraphy* of China, 2) *Carboniferous Boundaries of China,* and 3) *Bibliography of Carboniferous and Permian Stratigraphy and Palaeontology.* Special issues on Carboniferous-Permian palaeontology were also presented in the journals *Acta Palaeontologica Sinica* and *Acta Micropalaeontologica Sinica.*

The International Permanent Committee for the Carboniferous Congress consists of representatives from Argentina, Australia, Belgium, China, France, England, Germany, Japan, the Netherlands, Poland, Portugal, Spain, USA and USSR, including the Chairmen of the International Subcommission on Carboniferous Stratigraphy, IUGS, the International Commission on Paleozoic Microflora, and the International Committee for Coal Petrology. The first meeting was held in Nanjing in July, 1985. Highly valuable guidelines on planning the scientific programs of this Congress were offered at the meeting. On 30th of August and 3rd of September, 1987, the Committee held the second and the third meetings in Beijing. All of the members participated in these meetings, and only the representative of USSR was regretfully absent. Because the themes of the Congresses increasingly stress Permian stratigraphy and geology, the Committee decided at the third meeting that beginning with the XIIth Congress, a revised name, the "International Congress on Carboniferous-Permian Stratigraphy and Geology" should be used. The next congress will be held in the southern hemisphere (Argentina) with special attention to the problem of correlating Late Paleozoic strata of Gondwana and the Northern Hemisphere. The Committee accepted Dr. Gorden's suggestion of retirement and expressed appreciation for his great contributions to the Congresses, especially in organizing the IXth Congress in Washington D. C. and Urbana in 1979 as its President.

The International Conference Center for Sciences and Technology handled much of the behind-the-scenes planning and implementation for the XIth Session, including convention, housing, travel and bookkeeping services. Their professional assistance allow the Scientific Secretariat to focus all of its attention on the scientific aspects of the Congress.

Three information circulars were published. The first one, totalling 3 700 copies, was distributed in the fall of 1984. By the end of 1985, some six hundred persons had returned their replies to this circular. The Second Circular described in great detail the proposed activities and arrangements. In all, 300 copies were distributed in the Spring of 1986. The Final Circular was mailed to registrants in May, 1987. It included a timetable for the scientific program and supplements to the information contained in the Second Circular.

Field excursions

Planning for the field excursions of the XIth-ICC began more than four years before the Congress convened. Among twelve excursions proposed for the Congress, eight were selected by the Chinese Program Committee at its first meeting in 1985. Four pre-Congress excursions were designed mainly to show the Carboniferous-Permian coal-bearing deposits in North China, while four post-Congress excursions lay stress on the

Carboniferous-Permian carbonate deposits in South China. The Excursion to Shanxi, one of the most extensively studied areas, was run twice. At the request of the International Working Group on the Devonian/Carboniferous Boundary, a pre-Congress excursion to visit the candidate sections in Muhua of Guizhou and in Nanbiancun of Guangxi was specially arranged. A group of four participants attended this excursion which was guided by Wang Chenyuan. Also, during the Congress, a one-day excursion to visit the Beijing Ape-men Museum and the Carboniferous-Permian strata in Xishan had taken place. Some 14 participants joined in this excursion. Owing to insufficient registrants, the excursions to central Hunan and to central Henan were cancelled at the 2nd meeting of the Chinese Program Committee held in March, 1987.

All excursions, including those cancelled were well prepared and operated by their local hosts independently. In addition to guidebooks, the monographs revealing new palaeontological and stratigraphical data of the Carboniferous-Permian sections to be visited were published and presented to the Congress. The diverse excursions provided the means by which geologists from other countries could not only exchange view points on the geological phenomena they have observed but also to make acquaintance with local Chinese geologists and people. For most participants, the excursions provided the highlights and most memorable aspects of the XIth-ICC. In all, a total of 167 foreign registrants joined in the excursions. The excursion to Xinjiang may be counted as the most popular one. All 59 foreign registrants enjoyed the characteristic geology and landscape of northern Mts. Tianshan as well as the warm hospitality from the local people.

The following is a list of the hosts and guides of each field excursion and the publications related to it.

Excursion 1 Carboniferous and Permian stratigraphy in Shanxi

It was hosted by the Corporation of Exploration and Geology of Coal Fields, Shanxi Province, Xi'an Branch of Chinese Academy of Exploration and Geology of Coal Fields, and Nanjing Institute of Geology and Palaeontology, with Xu Huilong, Pan Suixian, Zhao Xiuqu, Hou Jiewei and Liu Loujun as guides. Publications include *Sedimentary Environments of Xishan Coalfield, Shanxi* and *Palaeontology and Stratigraphy of Late Paleozoic Coal-bearing Strata in Southeastern Shanxi.* Altogether, 23 participants took part in the pre-Congress trip and 20 in the post-Congress trip.

Excursion 2 Carboniferous and Permian coal basins of central Henan

It was cancelled. A monograph entitled *Palaeontology and Stratigraphy of Carboniferous and Early Permian* was published by the Geological Institute of Henan.

Excursion 3 Carboniferous stratigraphy in Jingyuan, Gansu

Its hosts include the University of Lanzhou and Nanjing Institute of Geology and Palaeontology with Shen Guanrong and Wang Zhihao as guides. This took place in conjunction with a visit to the mid-Carboniferous boundary candidate section by the International Working Group. A total of 23 participants attended the excursion.

Excursion 4 Carboniferous and Permian stratigraphy in eastern Mts. Tianshan

It was co-hosted by the Nanjing Institute of Geology and Palaeontology and the Xinjiang Branch, Chinese Academy of Sciences, guided by Liao Zuoting and Ouyang Shu, and had 59 participants.

Excursion 5 Carboniferous and Permian stratigraphy in Guizhou

It was hosted by the Geological Bureau of Guizhou Province and guided by Wei Jiayong, Wang Gang, Ge Ke, etc., and had 22 participants.

Excursion 6 Carboniferous carbonate sequences of Guangxi

It was hosted by the Geological Society of Guangxi, with the guides Zhao Shun, Kuan Guodun and Ruan Yiping, and had 45 participants.

Excursion 7 Carboniferous stratigraphy of central Hunan

A monograph entitled *Palaeontology and Stratigraphy of Late Devonian and Early Carboniferous in Hunan* and a guidebook published by the Regional Geological Survey Team of Hunan Province.

Excursion 8 Carboniferous and Permian stratigraphy of Jiangsu and Zhejiang

The hosts included the Nanjing Institute of Geology and Mineral Resources, Nanjing Institute of Geology and Palaeontology, and the Institute of Oil Geology of Zhejiang, and the guides were Xu Shanhong, Tang Yi and Wang Yichen. Sixteen participants joined in this excursion.

Scientific sessions

There were 402 attending members, of which 182, together with 42 accompanying members, came from 30 countries outside of China. These include Argentina, Australia, Belgium, Canada, Czechoslovakia, D.P.R.Korea, Egypt, England, France, F.R.Germany, D.R.Germany, India, Ireland, Israel, Italy, Japan, S. Korea, Madagascar, Malaysia, Mozambique, the Netherlands, New Zealand, Poland, Portugal, Romania, Spain, Sweden, Switzerland, USA, and USSR. There were 32 non-attending members registered.

Two volumes of abstracts, containing 244 items provided by Chinese geologists and 201 by geologists from other countries, were published and distributed to the registrants together with other documents for the Congress.

The inaugural ceremony was held in the People's Great Hall. Prof. Wu Heng, President of the XIth-ICC declared the opening of the Congress. In succession, addresses were delivered by Chinese State Councillor Mr. Kang Shi'en, Prof. C. Virgili on behalf of the Chairman of the Spanish Organizing Committee for the Xth-ICC, and Dr. J. W. Cowie as the representative of IUGS. In commemoration of the sixtieth anniversary of the ICC, Dr. W.H.C.Ramsbottom was invited to give a review on the history of the Congress. Finally, Drs. R.H.Wagner, C.F.Winkler-Prins and L.F.G.Grand presented the newly published books *the Carboniferous of the World* to the Chairman.

Following the inaugural ceremony was a plenary session held in Beijing Science Hall, where the Congress took place. The lectures planned to be given in this session were the briefings of the main aspects of the Carboniferous in China and summaries of scientific results of newly achieved important international program. These include:

— The Carboniferous System of China (Wu Wangshi)

— The Major subdivisions of the Carboniferous System (R. H. Wagner)

— The Tectono-palaeogeography and biogeography of China and adjacent regions in the Carboniferous Period (Wang Hongzhen)

— Applications of palaeomagnetism in the Carboniferous (D. H. Tarling)

— The exploration and exploitation of coal resources in China (Wang Zhongtang)

— Carboniferous paleogeographic development in central Europe (E.Paproth)

Further, a total of 244 papers, including 106 by Chinese geologists, were presented in 21 smaller sessions. The general subject matter for the Congress was divided into 8 sections and 11 symposia on topics of general interest or on those connected with the current task of international scientific organizations as shown in a list of Congress sections and symposia. A half-day colloquium of "Approach to a Global Stratigraphic Scheme" was organized. Dr. Cowie (Chairman of the International Commission on Stratigraphy), Dr. Remane (Secretary General of ICS) and Dr. Paproth (Chairwoman of International Working Group on Devonian/Carboniferous Boundary) were invited to give talks on this important international program from their own points of view.

In association with the XIth-ICC, the 40th Annual Meeting of the International Committee for Coal Petrology was held in Shangyuan Hotel, Beijing from August 25-29. Thirty-three registrants attend this meeting, of which 16 non-Chinese were from 12 countries. Afterwards, a majority of the registrants joined in the scientific sessions of the XIth-ICC. An international classification has been suggested by ICCP after many years discussion, based primarily on the ranks of coalification, types and facies. Three subcommissions are respectively engaged in coal nomenclature and standardization, and the geological and industrial applications of coal petrology. The scheme for bitumen nomenclature had been adopted previously, and the new discipline of organic petrology has been instituted.

In addition to the ICCP, seven international organizations held business meetings after the symposia on their subjects during the XIth-ICC. At the business meeting of the International Commission on Paleozoic Microflora, a worldwide zonal correlation of Paleozoic miospores was established, especially in China, Ireland, Poland and the Donetz Basin, though there remains a shortage of data on Namurian miospores. Therefore, it was decided to set up a working group to study this subject.

The International Working Group on the Mid-Carboniferous Boundary discussed three candidate sections at its business meeting. These were the Jingyuan section of Gansu and the Ludian section of Guizhou, China, and the Arrow Canyon section, USA. The International Working Group on the Carboniferous-Permian Boundary was reorganized in Beijing with Wu Wangshi (China) and B. I. Chuvashov (USSR) as Chairwoman and Vice-Chairman, respectively. As regards the stratigraphical level of the Carboniferous-Permian boundary, there was a wide difference of opinion. The emphasis was on sequences close to the boundary on platform and slope facies in South China and the western Urals.

The question of a global stratotype for the Devonian-Carboniferous boundary attracted much attention during the Congress. Boundary sequences in shallow-water platform, slope, and basin facies in South China were reported in detail in a symposium. At the business meeting of the working group, it was decided to vote on the candidate sections at a meeting to be held next year in Ireland.

Most of the participants at the meeting of the International Committee on the Upper Permian were in favour of taking the top of Kungurian as the base for the Upper Permian Series, which would then include stages: Ufimian, Kazanian, Dzhulfian and Changhsingian, although opinions differ as to whether the Permian System should be subdivided into two or three parts. Correlation charts prepared by working groups in Australia, China, Japan, USA, USSR, and southern Europe were available.

The closing session, presided over by the Secretary General of the Congress, was held on September 4 in the Beijing Science Hall. Chairmen of the IUGS Subcommission on Carboniferous Stratigraphy, the International Commission on Paleozoic Microflora and the International Committee for Coal Petrology delivered reports of their activities, respectively. The Chairman of ICCP presented the Thiessen Medal to Dr. D.G. Murchison, which his representative accepted for him as he couldn't attend the Congress for health reasons. Finally, representative of Argentina, the host country for the XIIth International Congress of Carboniferous-Permian Stratigraphy and Geology expressed his warm welcome to participants who would be meeting again in the southern hemisphere.

ACKNOWLEDGEMENTS

The Editors for *the Compte Rendu*, on behalf of the Chinese Executive Committee for the XIth-ICC, gratefully acknowledge the co-sponsorship of the China Association for Science and Technology, the Palaeontological Society of China, the Geological Society of China, Chinese Society of Coal and Chinese Society of Petroleum, and the finacial contributions from the National Natural Science Foundation of China, the Third World Academy of Sciences, the Assocication of Geologists for International Development, and the International Union of Geological Sciences. The organization benefitted greatly from collaboration of the Nanjing Institute of Geology and Palaeontology, Chinese Academy of Sciences, and the International Conference Center for Science and Technology, and from the enthusiastic cooperation of the individuals and organizations who prepared and conducted the field excursions, who made it possible to hold the XIth-ICC in Beijing, and who contributed to the publication of *the Compte Rendu*.

XIe Congrès International de Stratigraphie et de Géologie du Carbonifère Beijing 1987,
Compte Rendu 1 (1991): 8-9

Opening Address

Wu Heng

Our respectable members of the International Permanent Committee, our honourable delegates to the Congress, and our dear colleagues and guests:

In my capacity as chairman of this Congress, I now take great pleasure in announcing the inauguration of the XIth International Congress on Carboniferous Stratigraphy and Geology.

The International Congress on Carboniferous Stratigraphy and Geology has already had a history of 60 years since the First Congress was held in Heerlen, Holland in June, 1927 up to the XIth Congress held at this time. Carboniferous strata are well known for their abundant coal resources, together with such important natural resources as the combination of gas, bauxite, and iron ore, etc. In the past 60 years, the Congress has made great contributions to the promotion of prospecting and exploiting Carboniferous mineral products, and also the development of their related geological theories and some new branches of the geological sciences. All the previous Congresses were held in the developed countries in Europe and America; this is the first time the Congress is convened in a developing country. This Congress, I believe, will surely make new contributions to the promotion of economic development in the third world; it will become more prosperous than ever, and thus will be increasingly welcomed by the geologists of different countries over the world.

China is a country with a vast territory, in which Carboniferous strata are completely developed and varied in type. There is a current international academic activity aimed at establishing a global Carboniferous boundary stratotype. It is hoped to establish a competitive candidate section in China, whether the section be an uppermost, basal, or a medial boundary. In the Carboniferous and Permian of China, coal deposits account for more than one third of the national reserves, while more than 90% of its total bauxite reserves, are found in these beds. Between the Subsystems of the Upper and Lower Carboniferous there are important deposits of level-controlled iron ore, manganese ore and poly-mineral ores. Also, the Carboniferous strata serve as the raw-ore reserve beds of some combination gas fields. In addition, it is worthwhile to explore the genesis of many mineral resources and the regularity in their distribution.

You are heartily welcome to come to Beijing and to take part in this Congress, and you are also expected to initiate academic exchanges in many fields with Chinese scholars, whether on the spot or in the meeting rooms.

At the present time, this old country of ours is striving for the realization of modernization. You are welcome to have a look at Chinese ancient civilization, such as the Great Wall and Qinshihuang (the First Emperor of the Qin Dynasty) pottery figures and horses, and also the peculiar scenery, such as the mountains and rivers in Guilin, Guangxi, and the Gobi Desert in the Tianshan area of Xinjiang. Of course, you may also obtain a good feeling for the spirit and new look of the present age Chinese people. May this Congress become a bridge leading to the strengthening of friendship and understanding among the people of different countries.

Wishing this Congress a great success.

XIe Congrès International de Stratigraphie et de Géologie du Carbonifère Beijing 1987,
Compte Rendu 1 (1991): 10-11

CONGRATULATORY SPEECH

Kang Shi'en

Chairman, our respectable members of the International Permanent Committee, delegates to the Congress, guests and friends:

On the occasion of the inauguration of the XIth International Congress on Carboniferous Stratigraphy and Geology in Beijing today, and on behalf of the State Council of the People's Republic of China, I would like to extend our warm congratulations to the Congress and express our warm welcome to the geologists and specialists from nearly 30 countries and districts over the world!

It has been just 60 years since the 1st Congress held in Holland in 1927. All the previous congresses have organized experts of different nationalities and different disciplines in earth sciences and developed both global and regional scientific researches on the Carboniferous strata, thus making significant contributions to the exploration and exploitation of energy resources and minerals in different countries as well as to the establishment and development of the related disciplines.

The Carboniferous strata of China are extensively distributed and fully developed, rich in fossils and various in sedimentary types, with the extremely rich coal resources, oil, natural gas, and other metal or nonmetal mineral resources. During the past years, the Chinese geological workers have done a great deal geological exploration and scientific research work in this field. In particular, in recent years, overall conclusions and monographic researches have been made on the Carboniferous strata and the minerals contained therein by the application of modern theories and methods in geological science, with the achievement of valuable results. In the international activities to discuss the establishment of a global stratotype boundary section for the Carboniferous, the geological circles of China will surely make their contributions to the solution of this problem.

The previous International Carboniferous Congresses on Stratigraphy and Geology were held for ten times in those developed countries in Europe and America. But the current congress is now being held for the first time in one of the developing nations in the third world. This is also the first large-scale international symposium on earth sciences to be held in China. At the Congress, we'll discuss extensively all the problems on the Carboniferous paleoenvironment, the formation and distribution of minerals as well as stratigraphy. This is indeed a good chance for we Chinese scholars to learn more and exchange views with our colleagues from all over the world. At present, the Chinese Government is further pursuing the reform of government system and the realization of the open policy. The strengthening of international academic exchanges and the stepping up of the advanced science and technology is most important to the socialist and modernized construction of this country. It is my hope that the Chinese scholars at large will make use of this excellent opportunity to discuss problems of common interest with those scholars from other parts of the world. I also hope that after the congress, academic exchanges and cooperation between scholars in various countries will be further enhanced, so as to promote the development of geological sciences and the exploration of mineral resources in this country.

I firmly believe that, thanks to the opening of this congress, the friendship between scholars in various countries will be further strengthened and that this congress will make new contributions to the intensive studies of international Carboniferous stratigraphy and geology, and to the prosperous economy of various countries throughout the world.

Finally, I wish the congress a complete success! Thank you!

XIe Congrès International de Stratigraphie et de Géologie du Carbonifère Beijing 1987,
Compte Rendu 1 (1991): 12-15

Inaugural Address

Carmina Virgili

Distinguished authorities, Mr. Chairman, and the Chinese Organizing Committee of the XIth International Congress on Carboniferous Stratigraphy and Geology, members of the Honorary and Permanent Committee, friends and colleagues:

It is a great honour for me and gives me great pleasure to take part in the inaugural ceremony of this Congress and, with all our desires for success, to pass on the congratulations of the Spanish Organizing Committee of the Xth Congress. Mr. Adriano Garcia Laigoroy, who for personal reasons has not been able to accompany us, has specially asked me to convey his best wishes to you.

All of us who took part in the preparation of the Xth Congress are aware of the amount of work and dedication involved in the organizing of a Congress of this category, and we therefore particularly appreciate the efforts of the Organizing Committee of the XIth Congress being inaugurated here today.

It is now 60 years since a group of European coal-producing countries had the idea of meeting to exchange information on Carboniferous stratigraphy. Soon, countries in Asia, America and Australia joined the project, at the same time as the objectives were diversified. So, while the Subcommission on Carboniferous Stratigraphy meets, the Subcommission on Permian Stratigraphy, the Committee on Coal Petrology, and the Commission on Palaeozoic Microflora also meet.

In 1975, after the first seven Congresses were held in Central and Northern Europe, the VIIIth Congress took place in Moscow, and it was therefore possible to study the classical sections of the Soviet Union's Carboniferous and Permian.

In 1979, the IXth Congress was held outside Europe for the first time at Urbana in the U.S.A., enabling us to study the U.S. Carboniferous and Permian strata that significantly differ from European ones.

The Xth Congress once more brought us in 1983 to Europe, to the Iberian Microplate, located in the extreme southwestern part of Europe.

From a very early stage, that is since 1951—the date of the third Congress held in Heerlen—Spanish geologists had started taking part in the Congress on Carboniferous Stratigraphy and Geology. But it was not until 1963, at the Congress in Paris, that the Spanish participation became more intensive and its representation more considerable. This coincided with an increase in research work on our Carboniferous deposits.

Since the early 80's, the Spanish national energy policy has been one of clear support and development of coal production. This in turn has resulted not only in a deeper knowledge of those basins already known and exploited since ancient times (like those in Asturias, Leon or Penarroya...) but also in a search for new deposits, which in some instances has been completed with great success, as in the case of the lignites of Galicia.

This new interest in coal research was the main reason for which Spain offered to organize the Xth Congress. On that occasion, my country felt very much honoured with the visit of specialists from all parts of the world. For us it was of a great interest to visit with them at our outcrops and deposits and to be able to offer them the possibility of widening their knowledge of our country.

With the collaboration of Portugal, Spain prepared six geological excursions that enabled us to get to know the most important Carboniferous basins and the different facies of the Carboniferous of the Iberian Peninsula, as well as the most important mineralization processes linked to these materials.

Congress sessions took place from the 12th to the 17th of September. There were 8 sessions, 9 symposiums and 5 large working groups, attended by 405 participants and 80 other persons, either attending privately or representing Spanish organizations.

In the 8 sessions a total of 149 communications were presented, the majority referring to sections on:

1) Stratigraphy and Correlations

3) Palaeontology and Palaeoecology

6) Sedimentology and Sedimentary Environments

8) Geotectonics

The sessions on:

2) Economic Geology and Coal Prospection

4) Palaeogeography and Palaeoclimatology

5) Mineral Deposits — Excluding Coal

7) Coal Petrology

were less attended but provoked just as much interest.

Bearing in mind the subjects previously suggested by participants, 9 Symposiums took place:

1) "Subdivision, on a world-wide scale, of the Lower Carboniferous", for which 6 reports were presented.

2) "Change in sea level during the Carboniferous", for which 3 reports were pre-

sented.

3) "Global geography during the Carboniferous", for which 4 reports were presented.

4) "Economic geology: coal resources and exploration", for which 4 reports were presented.

5) "Upper Devonian-Carboniferous boundary" for which 6 reports were presented.

6) "Plate adjustment during the Carboniferous", for which 3 reports were presented.

7) "Cinderites and tonstein", for which 9 reports were presented.

8) "Gondwana Carboniferous and Lower Permian", for which 6 reports were presented.

9) "East Asian Carboniferous", a symposium we can consider as the forerunner of the Congress being inaugurated today.

In this brief review of the Congress's activities we should also mention the meetings held by several international subcommissions and working groups, for instance:

"International Subcommission on Carboniferous Stratigraphy".

"International Subcommission on Permian Stratigraphy: Carboniferous/Permian Boundary".

"Global Correlation of Carboniferous Formations".

"Working Group on the *Sphenopteris* Genus".

"International Working Group on Carboniferous and Permian Vegetable Compression".

Before the Congress, besides the guide books on the excursions, a volume was also published summarizing the 292 communications. A book on the Spanish Carboniferous and Permian and another on the Portuguese Carboniferous, were also printed, aimed at giving an overall view of the geological formations of these systems in the two countries.

While the Congress was in progress, the first of four volumes on *The World Carboniferous* made its appearance, sponsored by the International Subcommission on Carboniferous Stratigraphy and the Spanish Committe on Stratigraphy, and dedicated to the Chinese, Korean, Japanese and Southwest Asian Carboniferous.

In 1985 four more volumes were published with more than 1 500 pages, detailing the Congress's records and communications.

The Spanish Organizing Committee therefore considers that it has fulfilled its commitment of assuming the organization of the Xth Congress.

This XIth International Congress on Carboniferous Stratigraphy and Geology being inaugurated here in Beijing today is the first held on the Asian Continent.

It is taking place in a magnificent country, not only as regards the area it occupies but also as regards natural resources and, more especially, human resources.

This is a legendary country on account of its past, attractive as regards to the present, and fascinating looking to the future. It is a country in which ancient philosophical and artistic wisdom goes hand in hand with modern scientific and technological knowhow, a country geologists from all over the world have come to because of their interest in geological outcrops, and on account of the quality of the research carried out by Chinese scientists.

The carefully prepared geological excursions will enable us, or have already enabled us, to get to know the sedimentary environments and facies that are so varied and different, something that has never been possible to do in previous Congresses, until now, in any other country.

For this reason, the Madrid Congress gratefully accepted and looked forward to the offer by the Chinese People's Republic to organize the XIth Congress. Now that we have had the opportunity to observe the magnificent scientific proposals, organization of the Congress, and the warm welcome, we are even more satisfied and grateful.

Many thanks for everything and we wish the Congress every success.

XIe Congrès International de Stratigraphie et de Géologie du Carbonifère Beijing 1987, Compte Rendu 1 (1991): 16-19

Opening Speech

J. W. Cowie

The International Commission on Stratigraphy (ICS) of IUGS is responsible for the coordination of international stratigraphy, from the earliest part of the Archaean through the Proterozoic and the Phanerozoic: physical, chemical and biological aspects. A major current focus is the selection and definition of boundary stratotypes. Other branches of stratigraphy are of great importance too, and future changes and additions in our guidelines may well be required.

The most reliable systems of stratigraphy deal with global processes which are universal, unidirectional in the sense of irreversible — time sequence can only be read one way — and non-recurrent and non-repeatable. Included here, most significantly, is the evidence from biological evolution (sequential) and nuclear decay (metric). Biological evolution interacts through geological time with other factors, but is the main indicator of the direction of the arrow of time, of prime polarity. The evidence available so far shows that it cannot be stopped and reset. Nuclear decay also has this polarity, but unlike biological evolution it can be stopped and reset, and it has the great virtue of numeracy. Geochronometry has particular attraction for geoscientists working in unfossiliferous or sparsely fossiliferous rocks, but biostratigraphy gives the most useful and, at the present stage of research, generally the most accurate framework.

Today, stratigraphy is a subject in a dynamic phase of development with diverse emphases on aspects like unique or recurrent cyclic events (event stratigraphy), such as ash falls, eustatic changes, glacial deposits, appearance or disappearance of a particular biota, and evidence of impacts of extraterrestrial bodies. If these events can be shown to have global and isochronous effects, so that they are not merely parochial and diachronous masquerades, then they can be uniquely valuable in elucidating earth history. Cyclicity is still being sought in the modern search for the 'pulse of the earth'. Adjectives attached to stratigraphy proliferate, indicating renewed interest and involvement with stratigraphy as the keystone of the geological sciences — ecostratigraphy, seismic stratigraphy, chemostratigraphy, event stratigraphy, biostratigraphy, magnetostratigraphy, and others.

A great mass of scientific data, ideas and theories has now been put forward formally and informally in printed publications, newsletters and circulars by Commission bodies, and the production of common, agreed stratigraphical standards is evident and will escalate. Fundamental work in the geological sciences transgresses national boundaries, and agreements on an international basis are particularly valuable. However, in some cases disagreements can be a seminal factor and can accelerate progress. The coordinating and integrating function of the Commission itself (through its Secretariat/Bureau), may become even more desirable if used with discretion; plans are now aimed at developing this further in the next four years. Increased understanding of useful stratigraphical definition and subdivision on an agreed international basis and improved refinement and resolution in global correlation are the main threads running through this work, and the impetus continues to be maintained under the leadership of IUGS.

The rest of this decade and the 1990s promise to be a productive period for geoscience. Stratigraphy in all its manifold and fundamentally important aspects will be much involved and will make its vital contribution.

In February of this year the Chairman's Report to the IUGS Executive Committee contained news which I hope will be of some interest to this Congress.

1. The composition of the Bureau of ICS was completed in 1986 with the appointment as 2nd Vice-Chairman of Professor A.J. Boucot (Oregon State University, Corvallis, Oregon, U.S.A.), by negotiation between the ICS Chairman and U.S. geologists concerned with the 28th International Geological Congress (Washington D.C., U.S.A., 1989).

2. The Chairman of ICS attended as observer (i) the annual meeting of the IGCP Board in Paris, France in early February 1986 where he reported briefly on the work of ICS and the role of stratigraphy in IGCP, and (ii) in mid-February the first part of the Annual Executive Committee meeting of IUGS at the National Academy of Sciences in Washington D.C., U.S.A.

3. The *ICS Guidelines and Statutes* were published by the Commission on Stratigraphy in June 1986; the Statutes had been ratified by the IUGS Executive Committee in February 1986. This 14 page paper had been circulated in draft to ICS bodies' officers and others: no basic criticisms were made but comments on detailed points resulted in modifications. Grateful acknowledgment of help is due to K. Perch-Nielsen, V.V. Menner, F. Rögl, F.F. Steininger, K. Plumb, A. Yu. Rozanov, D.L. Dineley, J.M. Dickins, A. Zeiss and particularly for prolonged and voluminous correspondence with Amos Salvador on this and other matters.

Criteria briefly listed in the Guidelines for assessing a Phanerozoic Global Boundary Stratotype and Point IN ROCK are:

1) Correlation on a global scale

2) Completeness of exposure

3) Adequate thickness of sediments

4) Abundance and diversity of well-preserved fossils
5) Favourable facies for widespread correlation
6) Freedom from structural complication and metamorphism
7) Amenability to magnetostratigraphy and geochronometry
8) Accessibility and conservation
We are expected by IUGS Executive to follow these Guidelines.

4. The Chairman and Secretary-General of ICS (separately) attended meetings of ICS bodies and an IGCP Project in 1986:

1) Chairman (i) a workshop in May of the Working Group on the Precambrian Cambrian Boundary, jointly with the Subcommission on Cambrian Stratigraphy, in Uppsala, Sweden on "Taxonomy and biostratigraphy of the earliest skeletal fossils", (ii) a workshop of IGCP Project 216 ("Global biological events in earth history") on "Global Bio-events" in Göttingen Federal Republic of Germany, in May, (iii) a workshop on a proposed global sedimentary geology program held in June/July in Miami, Florida, U.S.A. under the sponsorship of the Society of Economic Paleontologists and Mineralogists (SEPM).

2) Secretary-General (i) a workshop in May in Aberdeen, Scotland, U.K. on quantitative stratigraphy. (ii) a conference in July on Permian-Triassic stratigraphy in Brescia, Italy.

5. Progress with publication of outstanding parts of the International Stratigraphic Lexicon has been:

1) Volume 1 fascicule 7b *Swiss Molasse* was published in 1987 in Switzerland by the Swiss Geological Commission and the National Hydrological and Geological Service.

2) Volume 1 part 3a, Section IX. *New Red Sandstone of England, Wales and Scotland*: a decision is awaited on publication.

6. The 1st (1976) edition of the *International Stratigraphic Guide* (*ISG*) is being revised and it is hoped that the second edition will further contribute to one of the objectives of ICS as stated in its Statutes (section 2a): "to clarify principles of stratigraphic procedure and unification of stratigraphic nomenclature". One or two Chairmen of ICS bodies have somewhat misunderstood the statement by the ICS officers (in a number of recent publications) that The *International Stratigrephic Guide* prepared and published by the Subcommission on Stratigraphic Classification of ICS, contains valuable discussion and recommendations but it was never adopted by ICS as a statutory policy document. After discussions following the publication of ISG in 1976 the Chairman of ICS circulated a statement in 1977 (after stressing the statutory policy of ICS given above), that the *Guide* itself stresses that it does not set out to be a *mandatory code*. This phrase "mandatory code" is perhaps a clearer equivalent of the phrase "statutory policy document" and is intended by the present ICS Bureau to have the same meaning. It should again be stressed that the principles laid down in the *Guide* allow for considerable adaptations of emphasis and structure and should continue to be tested so that a revised edition of the *Guide* can attempt to approach nearer to an international consensus without being emasculated.

7. Cooperation between the Chairman, Commission on Planetology (IUGS) and the Precambrian Subcommisson has commenced in the planetary side of early earth history such as impact features and general planetary evolution factors.

Agreement has been reached between the Chairman of ICS and the Chairman of the Precambrian Subcommisson to proceed with setting up an ICS Working Group under the Precambrian Subcommisson on the Terminal Precambrian System (Sinian/Vendian/Ediacaran) and a convenor is being nominated.

The first result is the publication and circulation of a *World Map of Impact Structures* by the Geological Survey of Canada circulated with the June 1987 issue of *Episodes*.

8. Devonian Subcommission

Submission has been made to ICS and agreed to by a large majority of the Commission's Voting Members on the Middle Upper series global boundary stratotype section and point. Ratification by IUGS Executive Committee will be sought in its Paris meeting in February 1987: granted in Paris on 7 February.

9. Cooperation has begun and a good response to a circular is being received between IUGS's Commission on Teaching of Geology and ICS (being organized by the Secretary-General of ICS) on the Teaching of Stratigraphy at all levels.

10. Global standard stratigraphic scale

Since the initiation of this project in 1984 by ICS Secretary-General M.G. Bassett at the 27th International Geological Congress in Moscow some progress has been made and cooperation continues between ICS and the IUGS Publications Board to produce and publish such a scale. A draft, at least, should be available before long and a publication could be available before the 28th International Geological Congress in Washington D.C., U.S.A. in 1989.

XIe Congrès International de Stratigraphie et de Géologie du Carbonifère Beijing 1987,
Compte Rendu 1 (1991): 20-27

A History of the Carboniferous Congress

W. H. C. Ramsbottom

INTRODUCTION

It is entirely appropriate that Carboniferous geologists should be gathering together in 1987, for this year is the bicentenary of the birth of W. D. Conybeare, the founder of the Carboniferous System. I shall start by saying a very few words about him, because it is unlikely that his name will be heard amongst us again during this week.

Conybeare was born in England in 1787, and by 1803 was a student of geology at Oxford University. When he was eighteen his grandmother died and left him an income of 1 500 pounds per year, equivalent to about 4 0000 pounds in todays money, and it was tax-free. He made a remarkable decision for so young a man. He found he could live comfortably and well in Oxford on one third of his income-500 pounds, so he decided to spend another third on books, and the remaining third on travel. How many of us would like to be able, or indeed be wise enough, to make a similar decision today? Travel in Europe was impossible because of the Napoleonic wars, so he travelled widely in England instead, and acquired an unrivalled knowledge of its geology. In 1822 he published a book on the geology of England and Wales in which he proposed the term Carboniferous in a way which is fully understandable today. Conybeare lived until 1857, but he took little interest in geology in the latter part of his life. His great achievement was to recognize the essential unity of the rocks now called Carboniferous, and these he divided into Carboniferous Limestone, Millstone Grit and Coal Measures (he also included the Old Red Sandstone, later removed as Devonian). Such was the influence of English geologists in the early nineteenth century that it is possible to find references to the terms Carboniferous Limestone, Millstone Grit and Coal Measures in, for example, both Russian and American literature until the nineteenth century was well advanced. But meanwhile other geologists in other countries have realised that, what was appropriate in Britain was not necessarily suitable in regions far away, and had proposed their own terms. In fact, by the early years of the twentieth century, a large number of stratigraphic terms had been applied to Carboniferous rocks in different parts of Europe, but there was little agreement as to their exact meaning or their correlation, and their definitions were frequently rather vague. All this is by way of background to the situation at the time of the First Carboniferous Congress held in 1927.

This is the Eleventh Carboniferous Congress and the first to be held in the eastern hemisphere. Your organising committee has therefore thought that it might be useful and appropriate to present a brief history of the background and achievements of the previous congresses, and of some associated bodies.

I have divided this account into three parts. First, the early congresses all held at Heerlen; then some accounts of the three independent though closely associated bodies which arose out of the activities of the congress; and lastly a somewhat brief account of the post-Heerlen meetings.

THE FIRST FOUR CONGRESSES AT HEERLEN

There is no doubt that the spiritual father and initiator of the Congress was Dr. W. J. Jongmans of Heerlen. Jongmans had been appointed Director of the Geological Institute at Heerlen, which is situated in the centre of the coal mining district of the Netherlands in 1921, and he was pursuing, in particular, the study of Carboniferous palaeobotany. He had many contacts among other Carboniferous workers, and he took advantage of a simultaneous visit to Heerlen in 1925 of Profs. Delepine and Pruvost (from France) and Prof. Gothan (from Germany) to suggest that an international meeting to discusss the problems of the stratigraphy of the Carboniferous might be useful. An initial organising committee comprising Jongmans, Gothan and Dr. A. Renier, from Belgium, was set up, and they sent out a preliminary circular and invitation in December 1926. The meeting was to be held at Heerlen under the auspices of the Geological and Mining Society of the Netherlands and with financial support from the Limburg coal mines.

The stated object of the meeting was to be "the comparison of the stratigraphy of the Carboniferous in the different coal regions of Europe, and to study the different methods of research leading to practical results in the various countries". The meeting was held in the mining school at Heerlen 11-27th June, 1927 and some 60 geologists attended. The Compte Rendue of the meeting was issued as a thick volume later the same year.

In order to appreciate the results of this First Congress it is necessary to know that, at that time, the notion of a stratigraphic classification in which time was separated from lithology was almost a new one. The great achievement of this meeting was to agree on a single classification of the West European Carboniferous which was generally accepted and was known as the "Heerlen Classification". What the congressists were in fact doing was to propose what we would now call a chronostratigraphy, though since that term had not been invented it was not specifically stated as such. The main new features of the classification were the recognition of a distinct and separate Namurian between the Dinantian and Westphalian, and the definition of Westphalian A, B and C. The principal underlying the classification was the recognition, resulting from the work of Herman Schmidt and W. S. Bisat that the goniatites were evolving so rapidly that they could provide a framework of precision time, in effect, on which to base the stratigraphy. All except the highest of the boundaries in this classification were based on the occurrences of goniatites.

The membership of this first Congress was almost all from Western Europe, and their aims were strictly limited to providing a classification applicable to that region.

Eight years later, in 1935, a second meeting was held, also at Heerlen and again organised by Jongmans, Gothan and Renier. Whereas the primary aim of the first Congress had been to organise the stratigraphy, this second meeting also added tectonics and coal petrography to the subjects discussed, and the stratigraphical importance of plant spores was also introduced. At this meeting, it was also attempted to correlate the Carboniferous successions in the U.S.A. and U.S.S.R. with those in Western Europe. The emphasis, however, was still on stratigraphy, and the Heerlen Classification was given added precision in certain of its features. The Dinantian was divided into three, the division of the Namurian in A, B, and C was made, and a formal Westphalian D was added to the Westphalian out of strata that had been considered hitherto to be of Stephanian age. Except for the Westphalian C/D and higher boundaries all the division were based on the occurrence of goniatites. There was considerable discussion of the boundaries between the Carboniferous and the Devonian (taken at the base of the *Gattendorfia* Stufe), and of the Carboniferous-Permian boundary.

In addition to the European representatives there were delegates from the U.S.A., the U.S.S.R., China, and India; the Congress was becoming more truly international.

The Third Congress was held in 1951, in the aftermath of the World War, and Jongmans was the Honorary President. Representatives from 14 countries attended. A major question at this meeting arose out of a request from the American Commission on Stratigraphic Nomenclature for the Congress to consider and give the opinion of its members on the advisability of dividing the Carboniferous System into two systems—Mississippian and Pennsylvanian, a matter which was also to be discussed at the International Geological Congress at Algiers in 1952. As a result of a public symposium on this subject, it was decided to recommend:

1. that the Carboniferous is a unity and shall remain classified as a System. Also that the System comprises an upper and a lower part which should be classified as "Sub-Systems" (but, see below, this decision was later rescinded).

2. since it was not yet possible to correlate the boundary between these two parts, no special name was proposed and the two Sub-Systems should be referred to as Upper and Lower Carboniferous.

North American delegates did not like this proposal and entered a caveat.

At this Congress also there were Round Table conferences on palynology and coal petrography. These were the precursors of the commissions on these topics which were later established. Applied geology was, for the first time, added to the official programme, and for the first time a field excursion was held, albeit to study the sedimentology of recent and subrecent sediments believed to be analogous to those of the Carboniferous.

The Comptes Rendus of the first two congresses had the somewhat cumbersome title *Congrès pour e' avancement des études de stratigraphic Carbonifère*. For the Third Congress the title was widened to include geology as well as stratigraphy. The word "international" was not included in the title until the report of the Fifth Congress in 1963.

The organization of the first three Carboniferous congresses, although sponsored by and funded from Dutch sources, had always remained a private matter arranged by *ad hoc* committees and not affiliated to any other organization. It was decided in 1951 that a properly organised international committee should be set up to organize any future meetings. The Permanent Committee (as it later became known) then appointed included representatives from the Netherlands, Belgium, Germany, France and Britain. Representatives from other countries were added in later years, but it has, to this day remained a private and independent organization. It may be metioned here that there was a proposal made at the Fifth Congress in Paris in 1962 that the Congress should become part of the International Union of Geological Sciences, but this was rejected by the Permanent Committee. This committee decided that although the Congress should in future meet in different countries, there should be one more meeting at Heerlen to coincide with the Fiftieth anniversary of the founding of the Geological Institute at Heerlen.

Accordingly the Fourth Congress was held at Heerlen in 1958, with participants from 19 different countries attending, the increasing economic importance of the Carboniferous coals was recognized in the programme of the meeting, and also by the fact that the High Authority of the Eupropean Coal and Steel Community provided financial support to supplement the local Dutch resources. Several excursions were held both before and after the Congress to regions of Carboniferous outcrops in the neighbouring countries. Some attention was paid on this occasion to hydrocarbons other than coal.

A stratigraphical colloquium was held, based on a questionaire on stratigraphic classification put to the members by Prof. Jongmans. The previous decision on the question of "Sub-Systems" was rejected, and the general opinion on the question of a single mondial classification was rather pessimistic. It was agreed, however, that the term Namurian should not be used in Russia in a sense different from that in Western Europe.

THE ASSOCIATED INDEPENDENT BODIES

Three new organizations grew out of the third and fourth Heerlen meetings. Although now independent organizations, they still retain a form of association with the mother congress by holding meetings at the same time and place as the main Congress.

The Fourth Congress saw the official creation of the Commission International de Microflore du Paleozoique (CIMP), with Prof. R. Potonie as Chairman and Dr. B. Alpern as Secretary. This Commission has always held meetings in association with the Carboniferous Congress at each subsequent Congress, and its Chairman is a member of the Permanent Committee. This Commission seeks to provide a zonation based on microfossils for the whole of the Palaeozoic, and it has been highly successful, at least as far as the Carboniferous is concerned. They issue a most useful newsletter. The current President is Dr. B. Owens, and the Secretary is Dr. G. Clayton of Ireland. There is a meeting of the Commission here in Beijing this week.

The International Commission on Coal Petrology (ICCP) also grew out of the meetings at Heerlen. At first concerned mainly with nomenclature, the Commission was later (in 1962, Paris) divided into two subcommissions, one on nomenclature with Prof. M. Teichmuller as President, and one on analysis with Prof. M.-Th. Mackowsky as President. This Commission too has held some of its meetings during the time of the Carboniferous congresses and its Chairman is also a member of the Permanent Committee. At the present time the President is Dr. B. Alpern, and the Secretary is Dr. Monica Wolfe of Krefeld, Germany. The Commission is now organized into three sections, one dealing with general matters of nomenclature and methodology, one on carbonisation, and one on geological features of dispersed organic matter. In spite of the title the field of study now includes all hydrocarbons, and not just coal. The ICCP has just concluded a highly succesful meeting here in Beijing.

The 1958 meeting in Heerlen also saw the amalgamation of the Temporary Commission on Carboniferous Stratigraphy set up in Algiers in 1952 as a result of the resolutions on stratigraphy of the third congress, with the International Subcommission on Carboniferous Stratigraphy (SCCS) of the Interntional Geological Congress. Dr. W.P.Van Leckwijck was the Chairman of the amalgamated body, which although meeting during the Carboniferous congresses is not formally associated with it, except that the Chairman of SCCS is an ex officio member of the Permanent Committee of the Congress, this arrangement allowing coordination of the activities of the two bodies.

The field of international stratigraphic classification, so successfully achieved at the first two congresses, has almost completely been taken over by SCCS. Resolutions on this topic now come, not from the Congress itself, as happened in earlier days, but from the Subcommission.

The work undertaken in the earlier years of the Subcommission was still primarily associated with elaborating the standard Heerlen classifications of 1927 and 1935. Its decisions on this topic were published in 1971, as a result of a discussion at the Congress in Krefeld. The main new feature then was the recognition, as a result of the work of Dr. R. H. Wagner of a Cantabrian Stage, between Westphalian D and Stephanian A. The latest work on the West European succession was done at Madrid in 1983, when names were given to the Westphalian A, B and C stages.

The work of establishing one single classification for Western Europe, undertaken at the first five or six congresses, was relatively easy compared with the attempt to achieve a worldwide classification, which is the current aim of the SCCS. The Carboniferous presents peculiar problems in this respect when compared with other systems. It is well appreciated that the European classification cannot serve as a globally applicable one, mainly for facies reasons in the upper part of the system. There are three or four well-entrenched classifications in use in various parts of the world, and there is no universal agreement, even now, on the usages of such basic terms as Upper and Lower Carboniferous, or even whether the Carboniferous should be divided into two or three major divisions. This topic, indeed, is to be under discussion this week. But a start has been made in the decision, taken by SCCS after the Madrid meeting in 1983 to seek a stratotype for a mid-Carboniferous boundary at the approximate level of the base of the *Homoceras* goniatite zone, the actual boundary being recognized by means of conodonts, however, rather than goniatites. This marks a departure, for the first time, to the principle, established at Heerlen in 1927 that divisions at least of the lower part of the Carboniferous should be based on goniatites, and is a result, of course, of the increased study of microfossils in recent years, and an appreciation that conodonts are frequently of more widespread occurrence, occurring in more different kinds of facies than do the goniatites. SCCS has a special Working Group, under the Chairmanship of Dr. H.R. Lane, which is seeking a suitable stratotype for this mid-Carboniferous boundary, intended to be globally applicable, and which is embarking on a series of field meetings to this end.

The general meetings of the Subcommission are open to all. It meets once every two years, and a general session is to be held here in Beijing this week. There are 29 voting members of the Subcommission, and a large number of corresponding members. At the moment I am the Chairman of the Subcommission, Dr. Robert Wagner is Vice-Chairman, and Prof. Walter Manger is the Secretary.

It ought to be made clear that the upper and lower boundaries of the Carboniferous are not the concern of this Subcommission, but of two separate and independent Working Groups of the Commission on Stratigraphy. The Working Group on the basal boundary, that is the Devonian /Carboniferous boundary, contains specialists from both systems and is under the chairmanship of Dr. E. Paproth. It has made good progress and has now selected a level, again, and for the same reasons, to be recognized on the occurrence of conodonts rather than by means of goniatites, which will form the base of the System, but has yet to select a single suitable definitive stratotype. This Working Group has made many excursions seeking such a place.

The Working Group on the Carboniferous/Permian Boundary has had a more chequered history and there is, alas, no progress to report.

THE POST-HEERLEN CONGRESSES

With the creation of the Permanent Committee and of the three other loosely associated but independent bodies (SCCS, CIMP and ICCP), at the last of the Heerlen meetings in 1958, the framework of future congresses was laid, and the story now becomes slightly different. I do not propose to go through in detail the work of each of these post-Heerlen meetings. These had now become like many modern scientific meetings, with a list of topics suggested by the local organizing committees, and a very large number of papers submitted and read. In addition, we now have some major symposia, often on fashionable topics of the day, and a few plenary lectures at the opening or closing sessions. The field excursions, less important at the earlier meetings, have now become an essential part of our proceedings.

With organizational aspects now in the background, attention became perhaps more concentrated on the geology, and on providing participants with an informed forum for presentation and discussion of their ideas and with opportunities to see for themselves on through the field excursions the Carboniferous geology of the different countries where the meetings were now to be held, and also to meet and establish personal contacts with geologists of these countries.

At first, attention was still fixed on Europe, with meetings in France (Paris, 1962), Britain (Sheffield, 1967) and Germany (Krefeld, 1971). Then it was in accord with the increasing internationalising of the Congress that the next two meetings should be in the U.S.S.R. (Moscow, 1975) and in the U.S.A. (Washington-Urbana, 1979), both these meetings opening up new vistas to many geologists, as much in the personalities met as in the magnificent field excursions provided. Next, Spain was visited in 1983, partly at least because of the belief that the Spanish successions would demonstrate some closeness between the Russian and European successions. The enjoyable excursions on this occasion proved that this was correct.

There remained one major area in the northern hemisphere which the Congress had not yet visited, and it was therefore with particular pleasure that the invitation from China was accepted by the Permanent Committee. Not only does this meeting allow an unusual opportunity to see the geology of China in person, but to make or renew many friendships.

The range of topics discussed at our meetings becomes ever wider. In the economic field papers on all aspects of coal still predominate, but those on oil and gas are increasing as time goes by. Palaeontology and biostratigraphy, together with the newer disciplines of palaeoecology and palaeoenvironments, are still prominent, but tectonics, sedimentology, geochemistry, and many other topics are all welcomed, and have been discussed in detail at the last few meetings.

The Permanent Committee now contains representatives from 19 countries, and may be said to be representative of nearly all the major countries where Carboniferous rocks are important. Attendances have increased, though slowly, and is naturally greater in the larger countries. In 1979 for example the membership was over 750, and more than 450 geologists went on the field trips.

Yet for all the good scientific discussions that have taken place at all the congresses, most people, I think, would say that it is not the science that they remember most when looking back on past congresses. Geology is advancing so rapidly today that an organization meeting only every four years, and with a rather slow publication schedule, is not likely to be the place where new ideas will be advanced. No, it is the friendships and good fellowship that linger in my mind the longest when thinking about any of the eight congresses that I have attended.

The Carboniferous Congress provides, above all, a wonderful opportunity to meet with and perhaps collaborate with, and perhaps engage in amicable arguments with friends from countries other than our own. I am quite sure that if this meeting lives up to its predecessors, before this Congress is over, all of us who are visiting your country will leave behind many friends and colleagues that we shall never forget, and with whom we shall keep in contact over the years ahead.

XIe Congrès International de Stratigraphie et de Géologie du Carbonifère Beijing 1987,
Compte Rendu 1 (1991): 28-29

Closing Address

Liu Dongsheng

Respectable Chairman,
Respectable delegates and friends:

The XIth International Congress of Carboniferous Stratigraphy and Geology is about to come to an end after 5 days of intense work.

At this Congress a total of 232 papers have been read, involving various aspects of stratigraphy, paleontology, and tectonic geology. For the host country, our Chinese colleagues have presented to the congress their summarized reports on the prospecting and exploration of Carboniferous strata, paleogeography, and coal resources, and their monographs on Carboniferous strata in such provinces as Shanxi, Guizhou, Hunan and Henan. They have also made public special issues and a series of publications as a token of the convening of the congress, so that the delegates from other parts of the world can have a better understanding of our research achievements and enter into extensive exchanges and cooperation with each other.

During the Congress, we have assembled a total of contributions, which we hope, will be put into publication as soon as possible, with several volumes to come out before the opening of the next congress.

During the specific panel meetings, different academic organizations from different countries lost no time in becoming positively engaged in cooperative research activities. We are pleased to learn that the new international Carboniferous-Permian Boundary Working Group has come into being, with Prof. Wu Wangshi and Dr. Chuvashov as leader and deputy-leader, respectively.

During the Congress, the International Permanent Committee for the ICC held two sessions with several important decisions being adopted.

The Permanent Committee feels gratful to Poland and Argentina for their enthusiastic invitation to become hosts of future Congresses. Since there is only one alternative, the Committee has decided that the next Congress will be held in Argentina, a country where the Gondwana strata are well-developed. We believe the next Congress will be an international gathering with distinguishing features.

We are also glad to find that the themes at the Congresses are increasing in number. Over the past several Congresses, the issues of stratigraphy and geology of the Permian have been taken as the main subjects of common interest. The study of coal geology is of even more importance. It has been therefore decided that the name of the congress should be changed to "International Congress of Carboniferous-Permian Stratigraphy and Geology".

We'd like to express our heartfelt thanks to Dr. Gorden of U.S.A., the ex-chairman of the IXth Congress, who has now retired from office as a permanent member after having made many contributions for the past decades to the Congresses.

There are new members to be added to the Committee. They are: Dr. Dutro of the U.S.A., Prof. Jin Yugan of China; Dr. Kato of Japan and Dr. Peryt of Poland.

May the Congress be more prosperous than ever before!

XIe Congrès International de Stratigraphie et de Géologie du Carbonifère Beijing 1987,
Compte Rendu 1 (1991): 30

Speech at the Banquet

Sun Honglie

Dear delegates to the Congress, and dear friends:

Upon reading a report entitled "Golden spike points to coal riches" by the journalist of the China Daily together with some other reports in the newspaper while staying outside Beijing, I felt very glad over the successful proceeding of the XIth International Congress on Carboniferous Stratigraphy and Geology. At the same time, I would like to express my warm congratulations on the prominent achievements as recorded in all the papers read out by geologists from different countries with regard to the elaborative reseaches on regional Carboniferous stratigraphy, the formation of a global frame of Carboniferous stratigraphy, and the exploration into the theories in such aspects as coal and other minerals.

In a series of different fields in geological sciences, it is necessary for us to make researches in a global foresight; on the other hand, international cooperation and academic exchanges are particularly important to studies in geology.

Before our departure, please kindly pass on to all other colleages my best wishes that scholars in defferent countries will make further contributions to the promotion of cooperative researches in earth sciences.

May all our participants make a successful geological excursion after the Congress!

May every one enjoy good health and have a good trip on your way home!

XIe Congrès International de Stratigraphie et de Géologie du Carbonifère Beijing 1987,
Compte Rendu 1 (1991): 31-40

Profile of Professor Duncan George Murchison — Receipient of Thiessen Medal

A. H. V. Smith

Despite the burden of administrative responsibilities imposed by his appointment as Pro-Vice-Chancellor of the University of Newcastle in 1986 Professor Murchison has declared his intention of continuing his teaching and research commitments in the Department of Organic Geochemistry. Such is the dedication of the man to the subject with which he has been associated all his working life. Duncan George Murchison graduated in Geology in 1952 and his interest in coal and petroleum owes much to his mentor Professor George Hickling who was already recognised as an authority on the botanical affinities of the woody components in coal based on the study of thin sections.

At this time the young Murchison was active in the life of his University both on the cricket field and in the Students Union Society of which he became President in 1953. In part, credit must go to Professor Erich Stach who, while visiting the Department, encouraged Murchison to follow his inclination and work for a time on coal. This he did with a research studentship from the National Coal Board but his interest in petroleum took him as a geologist to Royal Dutch Shell for a year after completing his PhD in 1957. Returning to his former college, later to become part of the University of Newcastle he embarked on a career which was to ensure his continued allegence to academia. In 1976 he was appointed to the Chair of Organic Petrology. It became increasingly difficult for him to devote time to research as he advanced up the academic ladder, first as Dean of the Faculty of Science and then as Head of the Department of Geology for four years prior to his present appointment.

The Department of Organic Geochemistry, with which his name has become synonymous, was formed in 1965 as part of the Department of Geology at Newcastle. Always a happy Department in which visitors are made welcome. Many of the younger generation of organic geochemists owe their success to the sound grounding received in the Department. International meetings often become occasions for the reunion of postgraduates who have established careers throughout the world.

From the beginning Murchison was quick to realise the potential of the newly emerging field of coal reflectance and his early researches were involved with the physical basis of reflectance of coals and other organic substances of high molecular weight. Latterly he has promoted research into the geothermal and tectonic influences on reflectance of organic matter in rocks. This involvement with the optical properties of coal can be recognised in a long series of publications, often in collaboration with his students. At the same time he appreciated the importance of combining microscopy with chemical and physicochemical properties of organic materials in order to better understand their nature and to characterise them. Early papers dealt with the nature and occurrence of macerals in bituminous coals and in particular highlighted the infra-red spectra of resinite and alginite and the changes that occur in chemical structure with coalification.

The receipt of various awards enabled him to develop and apply a combination of methods including optical, X-ray diffraction and electron microscopy to industrial problems. For the Steel Industry these included a study of the effects of using fresh and oxidised coals on the optical properties of cokes as well as the effects of varying the duration and rate of heating. For the Petroleum Industry investigations were made into the nature and distribution of organic matter in sediments in different depositional environments in relation to the occurrence of liquid hydrocarbons and natural gas. In all, Murchison has been author or co-author of over 50 publications, has contributed chapters to nine books or reviews and has acted as joint editor of three books.

Despite his academic commitments he finds time to participate in the activities of many outside bodies including the five British Geological Societies of which he is a member. He was an Executive Member of the British Mineralogical Society and of the Royal Microscopical Society and in recognition of his contribution to science he was elected Fellow of the Geological Society of London and of the Royal Society of Edinburgh and later was made an Honary Fellow of the Royal Microscopical Society. His association with the RMS as Chairman of the Materials Section (1972-1975) brought coal microscopy to the attention of a wider audience. Many will remember the successful symposium on Organic Sediments, Coals and Cokes, at Oxford in 1976, the year of his Presidency. Many of the presented papers were published as a special volume of the *Journal of Microscopy* in 1977. As Duncan remarked in the Introduction, the event marked the passage of organic petrology previously regarded as a somewhat esoteric activity to one with considerable implications for the fuel industries.

In the international field Duncan Murchison is one of the longest serving members of the International Committee for Coal Petrology and he soon became a respected member of the Executive for his clear grasp of essentials and ready ability to resolve protracted discussions that would otherwise have slowed progress towards a unified solution. He served the Nomemclature Committee as its Secretary and subsequently the whole

Committee as its Secretary Genearl before eventually becoming President in 1979. Since 1983 he has been the Treasurer in which capacity he has built up the finances of the Committee to a healthy position—something which he attributes to his Scottish ancestory.

If Duncan Murchison, like many Presidents of Student Unions, had entered politics his charisma would have taken him far. Fortunately, the academic world and not least, the ICCP, have benefited. Much liked by his friends and colleagues for his pleasant manner he has never adopted an aloof position with students that some in his position might have done. A hard worker himself he does not tolerate procrastination in others and his organising ability is valued and performed with a calm unhurried efficiency. His busy life leaves little time for hobbies but he is well known for being an excellent host and raconteur especially when good food and drink abound. Not so well known is his interest in a specialised branch of philately and his love of the open country and trout fishing which he finds a welcome relaxation from a hectic life.

Murchison is the first recipient of the Rheinhardt Thiessen Medal from the United Kingdom and it is a fitting reward for someone continuing the path initiated by Stopes, Hickling and Seyler all of whom used the microscope to great effect in their investigations of coal. The name of Duncan Murchison will surely continue to be associated with advances in the field of organic petrology and his many friends wish him every success in the future.

Publications List

Referred Papers

1. Reflectance of vitrinite.
   Brennstoff-Chemie 1957, 29, 47-50.

2. Upper-scale expansion in infra-red absorptionmetry.
   Chem. Ind. 1960, 1210.
   (with B. Sc. Crawford and R. E. Dodd).

3. The accuracy and the subjectivity factor of reflectance measurements with the Berek-microphotometer.
   Proc. 1st. Int. Congr. Coal Petrology 1960, 3, 49-57.

4. Polished surfaces of the coal macerals.
   Fuel Lond. 1961, 40, 398-406.
   (with E. H. Boult).

5. The term 'Micrite' in coal and limestone petrography.
   Fuel Lond. 1962, 41, 403-407.
   (with R. G. C. Bathurst).

6. Infra-red spectrum of resinite in bituminous coal.
   Nature 1963, 198, 254-255.

7. Properties of the coal macerals: elememtary composition of resinite.
   Fuel Lond. 1963, 42, 141-158.
   (with J. M. Jones).

8. The occurrence of resinite in bituminous coals.
   Econ. Geol. 1963, 58, 263-273.
   (with J. M. Jones).

9. Resinite in bituminous coals.
   In *Advances in Organic Geochemistry* 1963, 49-69 (Eds. Colombo and Hobson). Proc. Int. Meeting, Milan, 1962, Pergamon Press, Oxford, 330pp.
   (with J. M. Jones).

10. Optical properties of uranium oxides.
    Nature 1965, 205, 663-665.
    (with J. M. Jones).

11. Properties of coal macerals: infra-red spectra of resinite and their carbonized and oxidized products.
In *Coal Science* 1966, 307-331, Adv. Chem. Ser. No. 55, Amer. Chem. Soc., Washington D.C.

12. Biochemical alteration of exinites and the origin of some semifusinites.
Fuel Lond. 1966, 45, 407-415.
(with A. J. Bell).

13. Apparatus for reflectivity measurement on reactive and radioactive materials.
J. Microsc. 1968, 88, 503-512.
(with J. M. Jones, E.Scott and S. Pickles).

14. Small-scale drilling operations for research purposes.
Q. J. Engng Geol. 1968, 1, pt. 3, 195-205.
(with N. Farmer and J. M. Jones).

15. Properties of the coal macerals: infra-red spectra of alginites.
Fuel Lond. 1969, 48, 247-258.
(with R. Millais).

16. The identity of the Lickar limestone of Northumberland and its underlying coals.
Scott. J. Geol. 1970, 6, 200-207.
(with N. Farmer and J. M. Jones).

17. Identity of sporopollenins with older kerogens questioned.
Nature 1970, 227, 194-195.
(with B. Sc. Cooper).

18. Dispersion of the optical properties of carbonised vitrinites.
Fuel Lond. 1971, 50, 4-22.
(with R. J. Marshall).

19. Variation of vitrinite reflectivity in relation to lithology.
In *Advances in Organic Geochemistry* 1972, 601-612 (Eds. V. Gaertner and Wehner).
Proc. Int. Meeting, Hannover, 1971, Pergamon Press, Oxford, 736pp.
(with J. M. Jones and S. A. Saleh).

20. A British meta-anthracitic coal of Devonian age.
Geol. J. 1972, 8, pt. 1,83-94.
(with A. C. Cook and E. Scott).

21. Optically biaxial anthracitic vitrinites.
Fuel Lond. 1972, 51, 180-184.
(with A. C. Cook and E. Scott).

22. Optical properties of carbonized vitrinites--a reply.
Fuel Lond. 1972, 51, 252.
(with R. J. Marshall).

23. Optical properties of carbonized vitrinites.
Fuel Lond. 1972, 51, 322-328.
(with F. Goodarzi).

24. Oxidized vitrinites--their aromaticity, optical properties and possible detection.
Fuel Lond. 1973, 52, 90-92.
(with F. Goodarzi).

25. Optical properties of carbonized preoxidized vitrinites.
Fuel Lond. 1973, 52, 164-167.

26. Reflectivity and anisotropy of vitrinites in some coal scares from the Coal Measures of Northumberland.
Proc. Yorks Geol. Soc. 1973, 39, pt. 4, 515-526.
(with J. M. Jones and S. A. Saleh).

27. Petrology and rank of coals of the Splimersford borehole, East Lothian, Scotland.
Bull. Geol. Surv. G. B. 1974, 45, 99-111.
(with J. M. Jones).

28. Organic geochemistry of thermally metamorphosed fossil wood.
Fuel Lond. 1975, 54, 283-287.
(with J. Allan, E. Scott and S. Watson).

29. Resinite--its infra-red spectrum and coalification pattern.
Fuel Lond. 1976, 55, 79-83.

30. Petrography and anisotropy of carbonized preoxidized coal.
Fuel Lond. 1976, 55, 141-147.
(with F. Goodarzi).

31. Plant-cell structures in vitrinite chars.
J. Microsc. 1976, 106, 49-58.
(with F. Goodarzi).

32. The accuracy of refractive and absorptive indices from reflectance measurements on low-reflecting materials.
J. Microsc. 1976, 109, pt. 1, 29-40.
(with A. C. Cook).

33. A note on spherical structures in chars from vitrinites of coking rank.
J. Microsc. 1976, 109, pt. 1, 159-163.
(with F. Goodarzi).

34. Optical properties of graphite.
J. Microsc. 1977, 109, pt. 3, 289-302.
(with B. Kwiecinska and E. Scott).

35. Effect of prolonged heating on the optical properties of vitrinite.
Fuel Lond. 1977, 56, 89-96.
(with F. Goodarzi).

36. Thermally metamorphosed bitumen from Windy Knoll, Derbyshire, England.
Chem. Geol. 1978, 22, 91-105.
(with G. Khavari-Khorasani).

37. Influence of heating-rate variation on the anisotropy of carbonized vitrinites.
Fuel Lond. 1978, 57, 273-284.
(with F. Goodarzi).

38. Refractive index and absorption coefficient as measures of structural organization in carbonized bitumen.
J. Microsc. 1978, 114, pt. 2, 199-204.
(with G. Khavari-Khorasani and H. E. Blayden).

39. Microscope photometry in studies of the molecular structure of carbonized bitumens and pyrobitumens.
J. Microsc. 1979, 116, pt. 3, 337-349.
(with G. Khavari-Khorasani and H. E. Blayden).

40. The nature of Karelian shungite.
Chem. Geol. 1979, 26, 165-182.
(with G. Khavari-Khorasani).

41. Provincialism and correlations between some properties of vitrinites.
Int. J. Coal Geol. 1984, 3, 315-331.

(with J. M. Jones, A. Davis, A. C. Cook and E. Scott).

42. Optical properties of organic matter in relation to thermal gradients and structural deformation.
Phil. Trans. R. Soc. Lond, A 1985, 315, 157-186.
(with A. C. Cook and A. C. Raymond).

43. Initial vitrinite reflectance results from the Carboniferous of north Devon and north Cornwall.
Proc. Ussher Soc. 1987, 6, 461-467.
(with C. Cornford and L. Yarnell).

44. Retention of botanical structure in anthracitic vitrinites carbonized at high temperatures.
In press *Fuel Lond.* 1988.
(with F. Goodarzi).

45. Order of generation of petroleum hydrocarbons from liptinitic macerals with increasing thermal maturity.
In press *Fuel Lond.* 1988.
(with G. Khavari-Khorasani).

46. Lower Carboniferous coal depositional environments on Spitzbergen, Svalbard.
In press *Organic Geochemistry* 1988.
(with W. Abdullah, J. M. Jones, N. Telnaes and J. Gjelberg).

47. The influence of microbial degradation and volcanic activity on a Carboniferous wood.
Under review *Organic Geochemistry*.
(with A. C. Raymond, S. Y. Liu and G. H. Taylor).

48. Organic maturation and its timing in sedimentary kerogens: the Rashiehill borehole, Midland Valley of Scotland.
Under review *Scott. J. Geol.*
(with A. C. Raymond).

49. Effect of volcanic activity on level of organic maturation in Carboniferous rocks of East Fife, Midland Valley of Scotland.
Under review *Fuel Lond.*
(with A. C. Raymond).

Books

1. *Coal and Coal-bearing Strata.* Oliver and Boyd, Edinburgh, 1968, 418pp, 41pls, 127 figs. (Editor with T. S. Westoll).

2. *Microscopy of Organic Sediments, Coals and Cokes: Methods and Applications.* Proceedings of the meeting held in Oxford, 1976. *J. Microsc.* 1977, 109, 1-167.
(Organizer and Joint Editor).

3. *Microscope Photometry.* Proceedings of the meeting held in London,1978. *J. Microsc.* 1979, 116, 293-399.
(Organizer and Joint Editor).

Chapters in Books and Reviews

1. Reflectance techniques in coal petrology and their possible application in ore mineralogy.
Trans. Inst. Min, Metall. 1963-4, 73, 479-502.

2. Some recent advances in coal petrology.
C. R. 6e Congr. Intern. Strat. Geol. Carbonif. Sheffield (1967), 1969, 1, 351-368.

3. Organic geochemistry of coal.
In *Organic Geochemistry--methods and Results,* 1969, 699-726 (Eds. Eglinton and Murphy). Springer-Verlag, Heidelberg, 828pp.
(with B. S.Cooper).

4. The petrology and geochemistry of sporinite.
In *Sporopollenin,* 1971, 545-568 (Eds. Brooks, Grant, Muir, Van Gijzel and Shaw). Proc. Int. Symp., London, 1970. Academic Press, London. 718pp.
(with B. S. Cooper).

5. Coal.
In *The Planet Earth,* 1976, 153-157. (Ed. A. Hallam). Elsevier/Phaidon, 1977.

6. Optical properties of carbonized vitrinites.
In *Analytical Methods for Coal and Coal Products,* Ch. 31, Vol. II, 415-464. Academic Press, 1978.

7. Butcher, baker, candlestickmaker, but organic petrologists all. *Proc. Roy. Microsc. Soc.* (Presidential Address), 1979.

8. The energy gap--is coal the salvation?

Pap. Dep. Geol. Univ. Qd. 1980, 9, 2, 1-32.

9. Recent advances in organic petrology and organic geochemistry: an overview with some reference to "oil from coal".
In *Coal and Coal-Bearing Strata: Recent Advances* (Ed. A. C. Scott). *Geol. Soc. Sp. Publ.* 32, 257-302.

XIe Congrès International de Stratigraphie et de Géologie du Carbonifère Beijing 1987,
Compte Rendu 1 (1991): 41-43

Honorary Committee

Adriano Garcia-Loygorri (Chairman of X-ICC)
Huang Jiqing (Huang T. K.) (Honorary President of the Chinese Academy of Geological Sciences)
Kang Shi'en (State Councilor of the State Council, P. R. China)
Sun Honglie (Vice-President of the Chinese Academy of Sciences)
Tang Aoqing (National Natural Science Foundation of China)
Wu Heng (Chairman of the All-China Commission on Stratigraphy)
Wang Tao (Minister of Petroleum Industry, P. R. China)
Yu Hong'en (Minister of Coal Industry, P. R. China)
Zhu Xun (Minister of Geology and Mineral Resources, P. R. China)

Permanent International Committee for ICC

B. Alpern (I. C. C. P.)
S. Archangelsky (Argentina)
J. Bouckaert (Belgium)
Cheng Yuqi (P. R. China)
A. Garcia-Loygorri (Spain)
M. Gorden, Jr. (U. S. A.)
Jin Yugan (P. R. China)
K. H. Josten (FR Germany)
M. Kato (Japan)
Th. F. Krans (The Netherlands)
J. P. Laveine (France)
Liu Dongsheng (P. R. China)
J. T. Oliveira (Portugal)
M. G. Ortuno Aznar (Spain)
B. Owens (C. I. M. P.)
T. M. Peryt (Poland)
W. H. C. Ramsbottom (U. K.)
J. Roberts (Australia)
P. P. Timofeev (U. S. S. R.)
R. H. Wagner (S. C. C. S.)
Yang Jingzhi (P. R. China)

Chinese Organizing Committee

Chairman Cheng Yuqi
Secretary General Liu Dongsheng
Scientific Secretary Jin Yugan
Members Gao Lianda, Li Xingxue, Sheng Jinzhang, Tian Zaiyi, Wu Wangshi, Yuan Renguang, Yuan Yaoting

Chinese Program Committee

Chairman Yang Jingzhi
Members Chen Minjuan, Hou Hongfei, Li Yingpei, Mu Xinann, Pan Suixian, Yang Zunyi, Yang Shifu, Zhao Xiuhu, Zhang Linxin

XIe Congrès International de Stratigraphie et de Géologie du Carbonifère Beijing 1987,
Compte Rendu 1 (1991): 44-46

Participants of 40th ICCP

Prof. A. Aihara
Kyushu University 33
Dept. of Geology, Faculty of Science
Hakuoka 812
Japan

Dr. I. Bolkova
W. S. E. G. E. I.
Laboratoriia Petrographifuglia
74 Sredny Prospect
199026 Leningrad V-26
USSR

Prof. Dr. Boris Alpern
Gr. des Combustibles Fossiles
Geologie Appliquee
Universite D'Orleans
F-45045 Orleans Cedex
France

Dr. Margaretha Bengtsson
Studsvik-Energi-teknik AB
Blacksvampv 16
S-61163 Nykoeping
Sweden

Prof. Philippa M. Black
Dept. of Geology
University of Aucklard
Private bag
Auckland
New Zealand

Prof. Dr. A. C. Cook
University of Wollongong
Northfields Ave.
Wollongong
N. S. W. 2500 Australia

Prof. A. Davis
Coal Res. Sect., Dept. Geoscience
The Pennsylvania State University
513 Deike Building
University Park Pennsylvania 1680
USA

Dong Mingshan
South Coal Measuring Center
Ministry of Geology and Mineral
Resources
Guangzhou, China

Gong Zhichong
Institute of Geology and Exploration
CCMRI, Ministry of Coal Industry
Xi'an, China

Guo Mingtai
Dept. of Gelogy
Mineral College of Shanxi
Taiyuan, China

Dr. W. Hiltmann
Budnesanstalt fuer Geowissenschaften
und Rohstoffe
D-3000 Hannover 51
FRG

Jin Kuili
Beijing Graduate School

China Institute of Mining and Technology
Beijing, China

Dr. R. Kutzner
Gaertnerstrasse 3
D-4300 Essen 1
FRG

Li Yuxiu
Institute of Petroleum Exploratin and Exploitation
Beijing, China

Lu Jie
Institute of Geology and Exploration
CCMRI, Ministry of Coal Industry
Xi'an, China

Dr. Paul C. Lyons
Geological Survey,
956 National Center,
Reston, Va 22092
USA

Dr. Karl Ottenjann
Geologisches Landesant NW
de Greiff-Strasse 195
D-4150 Krefeld
FRG

Pang Zhijui
Beijing Graduate School
Wuhan College of Geology
Beijing, China

Ren Deyi
Beijing Graduate School
China Institute of Mining and Technology
Beijing, China

Dr. A. H. V. Smith
Yorkshire Regional Laboratory
National Coal Board,
Wath upon Dearne
Rotherham
South Yorkshire
United Kingdom

Dr. Y. Somers
INIEX
Rue de chera 200
B-4000 Liege
Belgium

Dr. Monika Steller
Berghau-Forschung GmbH
Franz-Fischer-Weg 61
D-4300 Essen 13
FRG

Tang Xiuyi
Huainan Institute of Mining and Technology
Huainan, China

Wang Chang
Institute of Geology and Exploration
CCMRI, Ministry of Coal Industry
Xi'an, China

Wang Jie
Coal High Education Section
China Institute of Mining and Technology
Xuzhou, China

Prof. Dr. Monika Wolf
Geologie, Geochemie u Langerstatten des Erdols u der Kohle
RWTR Archen
5100 Archen
Lochnerstrasse 4-20

Wu Jun
Beijing Graduate School

China Institute of Mining and Technology
Beijing, China

Yang Qi
Beijing Graduate School
Wuhan College of Geology
Beijing, China

Zhang Xiuyi
Institute of Geology and Exploration
CCMRI, Ministry of Coal Industry
Xi'an, China

Zhao Shiqing
Huainan Institute of Mining and Technology
Huainan, China

Zhou Shiyong
Dept. of Chemical Industry
Anshan Institute of Steel and Iron Technology
Anshan, Liaoning
China

XIe Congrès International de Stratigraphie et de Géologie du Carbonifère Beijing 1987,
Compte Rendu 1 (1991): 47

Invited Members

Chen Dun (Vice-Minister of Coal Ministry, P. R. China)
I. W. Cowie (The International Union of Geological Sciences)
Fang Jun (Head of Department of International Affairs, China Association for Science and Technology)
Hou Xianglin (President of Chinese Petroleum Society)
Lu Jiaxi (Vice-Chairman of China Association of Science and Technology)
Lu Yanhao (President of Palaeontology Society of China)
Niu Xijin (Secretary General of Chinese Society of Coal)
J. R. Remane (The International Commission on Stratigraphy)
Sun Shu (Head of Bureau of Resources and Environment, Academia Sinica)
Wang Ren (Vice-President of National Natural Science Foundation of China)
Wu Ganmei (Executive Director of China International Conference Center for Science and Technology)
Yan Dunshi (Chief Engineer of Ministry of Petroleum Industry, P. R. China)
Ye Lianjun (President of Chinese Sedimentology Society)
Zhou Mingzhen (Vice-President of Palaeontology Society of China)

XIe Congrès International de Stratigraphie et de Géologie du Carbonifère Beijing 1987,
Compte Rendu 1 (1991): 48-74

Attending Members

CHINA (201)

Bai Shunliang
Dept. of Geology
Beijing University
Beijing
China

Bi Wanchang
Nei Mongol Institute of Geology
Hulun South Road, Hohhot
Nei Mongol
China

Cai Tuci
First Regional Geological Survey
Team of Xinjiang
Qitai, Xinjiang
China

Chai Zhifang
Research Dept. of Nuclear Technological
Application,
Institute of High-energy Physics
Academia Sinica
P.O. Box 2732, Beijing
China

Chen Anning
Research Institute of Petroleum Exploration and Development, Changqing Oilfield
Qingyang, Gansu
China

Chen Jianguo
Yichang Institute of Geology and Mineral
Resources
P.O. Box 502, Yichang, Hubei
China

Chen Gengbao
Dept. of Geology,
Kunming College of Engineering
Kunming, Yunnan
China

Chen Peng
Beijing Research Institute of Coal
Chemistry CCMRI
Hepingli, Beijing
China

Chen Wenyi
Institute of Guizhou Bureau
of Geology and Mineral Resources
Beijing Road, Guiyang
Guizhou
China

Chen Xiaowei
Institute of Geology and Exploration,
CCMRI, Ministry of Coal Industry
Xi'an, Shannxi
China

Chen Yanyun
Dept. of Science and Technology,
Ministry of Geology and Mineral
Resources
Xisi, Beijing
China

Chen Zongqing
Sichuan Institute of Petroleum
Geological Exploration and
Development, Ministry of
Petroleum Industry
Yihaoqiao, Chengdu
Sichuan
China

Cheng Yuqi
Ministry of Geology and Mineral
Resources
Xisi, Beijing
China

Ding Hui
Dept. of Geology,
Shanxi Mining College
Taiyuan, Shanxi
China

Ding Peizhen
Xi'an Institute of Geology and
Mineral Resources
Friendship East Road, Xi'an
Shaanxi
China

Duan Shuying
Institute of Botany,
Academia Sinica
Beijing
China

Fan Yingnian
Chengdu Institute of Geology and
Mineral Resources
Renmin North Road, Chengdu
Sichuan
China

Fang Shaoxian
Southwestern Petroleum Institute
Nanchong, Sichuan
China

Feng Baohua
Institute of Mineral Deposits,
Chinese Academy of Geological
Sciences
Beijing
China

Gao Lianda
Institute of Geology,
Chinese Academy of Geological Sciences
Beijing
China

Geng Guocang
Institete of Exploration and
Development, Changqing Oilfield
Gansu
China

Gong Jingfu
Hunan Bureau of Geology and
Mineral Resources
9 Renmin Road, Changsha
Hunan
Chian

Gu Daoyuan
Jianghan College of Petroleum
Jiangling, Hubei
China

Guan Shiqiao
Geological Bureau, Ministry of Coal
Industry
Zhuozhou, Hebei
China

Gu Feng
Shenyang Institute of Geology and
Mineral Resources
Beiling Street, Shenyang
Liaoning
China

Guo Hongjun
Dept. of Geology,
Changchun College of Geology
Changchun, Jilin
China

Guo Shengzhe
Shenyang Institute of Geology and
Mineral Resources
Beiling Street, Shenyang
Liaoning
China

Guo Xinian
Henan Company of Coalfield Geology
Songshan Road, Zhengzhou
Henan
China

Han Dexin
Beijing Graduate School,
China Institute of Mining and Technology
Beijing
China

Han Shuhe
Dept. of Geology,
Northeast University of Technology
Nanhu, Shenyang
Liaoning
China

Hao Yichun
Beijng Graduate School,
Wuhan College of Geology
Beijing
China

He Kaishan
Hunan Regional Geological Survey Team
Jiangshe, Xiangtan
Hunan
China

He Xilin
Dept. of Geology,
China Institute of Mining and
Technology
Xuzhou, Jiangsu
China

Hou Jihui
114 Team of Shanxi Coal Geology and
Exploration Corporation
Daqing North Road, Changzhi
Shanni
China

Hu Shizhong
Nanjing Institute of Geology and
Mineral Resources, Ministry of Geology
and Mineral Resources
534 East Zhongshan Road, Nanjing
China

Hu Shouyong
Bureau of Natural Resources and
Environment, Academia Sinica
Sanlihe, Beijing
China

Hu Yicheng
Dept. of Geology,
Wuhan College of Geology
Yujiashan, Wuchang
Hubei
China

Hu Yufan
Paleobotanical Section,
Institute of Botany,
Academia Sinica
Xiangshan, Beijing
China

Huang Benhong
Shenyang Institute of Geology and
Mineral Resources
Beiling Street, Shenyang
Liaoning
China

Huang Caoming
Dept. of Geology,
Shanxi Mining College
Taiyuan, Shanxi
China

Huang Jixiang
Southwestern Petroleum Institute
Nanchong, Sichuan
China

Huang Zehui
Zhejiang Institute of Geology
Wulinmen, Hangzhou
China

Huang Zhixun
Chengdu College of Geology
Chengdu, Sichuan
China

Huo Fuchen
Bureau of Geology and Mineral Resources
of Ningxia
Yinchuan, Ningxia
China

Ji Chenglong
Shaanxi Company of Coalfield Geology
and Prospecting
Taiyi Road, Xi'an
Shaanxi
China

Ji Qiang
Institute of Geology,
Chinese Academy of Geological Sciences
Beijing
China

Jia Bingwen
Dept. of Geology,
Shanxi Mining College
Taiyuan, Shanxi
China

Jian Renchu
Dept. of Geology and Mineral Resources,
Ministry of Geology and Mineral
Resources
Beijing
China

Jin Kuili
Beijing Graduate School,
China Institute of Mining and
Technology
Beijing
China

Jin Yugan
Nanjing Institute of Geology and
Palaeontology, Academia Sinica
Nanjing
China

Jin Yun
Information Institute of Coal
Science and Technology

Hepingli, Beijing
China

Kuang Guodun
Institute of Guangxi Brueau of
Geology and Mineral Resources
1 Jianzheng Road, Nanning
Guangxi China

Lin Jinhua
CCMI, Hepingli, Beijing
China

Li Shegao
Dept. of Geology,
Xinjiang College of Engineering
Youhao Road, Urumqi
Xinjiang
China

Li Shoujun
Geological Exploration Dept.,
Oil College of East China
Dongying, Shandong
China

Li Wancheng
Henan Company of Coalfield Geology
Songshan Road, Zhengzhou
Henan
China

Li Wenguo
No. 1 Geological Regional Team of
Nei Mongol
Hohhot, Nei Mongol
China

Li Xingxue
Nanjing Institute of Geology and
Palaeontology, Academia Sinica
Nanjing
China

Liang Dingyi
Beijing Graduate School,
Wuhan College of Geology
Beijing
China

Liang Xiluo
Nanjing Institute of Geology and
Palaeontology, Academia Sinica
Nanjing
China

Liao Keguang
Institute of Geology and Exploration,
CCMRI, Ministry of Coal Industry
Xi'an, Shaanxi
China

Liao Zhuoting
Nanjing Institute of Geology and
Palaeontology, Academia Sinica
Nanjing
China

Lin Fansheng
Jiangsu Company of Coalfield Geology
and Exploration
Changzhou, Jiangsu
China

Lin Guowei
Zhejiang Institute of Oil Geology
Hangzhou, Zhejiang
China

Lin Jinlu
Institute of Geology, Acadmia Sinica
Beijing
China

Lin Yindang
Dept. of Geology,
Changchun College of Geology
Changchun, Jilin
China

Liu Diansheng
No. 1 Regional Geological Survey
Team of Heilongjiang
Xiaoling, Acheng
Heilongjiang
China

Liu Dondsheng
Instituute of Geology, Academia Sinica
Beijing
China

Liu Fa
Dept. of Geology,
Changchun College of Geology
Changchun, Jilin
China

Liu Huanjie
Dept. of Geology,
China Institute of Mining and
Technology
Xuzhou, Jiangsu
China

Liu Guanghua
Beijing Graduate School,
Wuhan College of Geology
Beijing
China

Liu Zuhan
Xiangtan Mining Institute
North Suburb of Xiangtan
Hunan
China

Lu Jie
Institute of Geology and Exploration,
CCMRI, Ministry of Coal Industry
Xi'an, Shaanxi
China

Lu Linhuang
Nanjing Institute of Geology and
Palaeontology, Acadmia Sinica
Nanjing
China

Lu Zhongji
Shandong Coal Geology Company
Jinan, Shandong
China

Lü Daosheng
Shanxi Company of Coalfield Geology
and Exploration
Xi'an, Shaanxi
China

Luo Jinding
Dept. of Geology,
Fuzhou University
Fuzhou, Fujian
China

Ma Gaoshang
Hebei Coal Geology Company
Hebei
China

Ma Xiaoda
Geological Research Institute of
Jiangxi
Xiangtang, Nanchang
Jiangxi
China

Mao Bangzhuo
Bureau of Geology,
Ministry of Coal Industry
Fanyang Road, Zhuozhou
Hebei
China

Mei Meitang
Beijing Graduate School,
China Institute of Mining and Technology
Beijing
China

Meng Hui
Bureau of Natural Resources and
Environment, Academia Sinica
Beijing
China

Mu Xinan
Nanjing Institute of Geology and
Palaeontology, Academia Sinica
Nanjing
China

Nan Yi
Institute of Geological Science of
Guangdong
Guangzhou, Guangdong
China

Ouyang Shu
Nanjing Institute of Geology and
Palaeontology, Academia Sinica
Nanjing
China

Pan Jiang
Geology Museum
Beijing
China

Pan Suixian
Institute of Geology and Exploration,
CCMRI ,Ministry of Cool Industry
Xi'an, Shaanxi
China

Pang Qiqing
Dept. of Geology,
Hebei College of Geology
Xuanhua, Hebei
China

Peng Zhengzhong
Research Institute of Petroleum
Exploration and Development,
Sichuan Petroleum Administration
Chengdu, Sichuan
China

Qin Guorong
Institute of Geological Sciences of
Guangdong
Guangzhou, Guangdong
China

Qin Subao
Changsha Institute of Geotectonics,
Academia Sinica
Tongzipo, Changsha
Hunan
China

Qu Weimin
Anhui Coal Geology Company
Hefei, Anhui
China

Ruan Yiping
Nanjing Institute of Geology and
Palaeontology, Academia Sinica
Nanjing
China

Rui Lin
Nanjing Institute of Geology and
Palaeontology, Academia Sinica
Nanjing
China

Shang Guangxiong
No. 1 Geological Prospecting
Company, Ministry of Coal
Industry
Handan, Hebei
China

Shen Guanglong
Dept. of Geology,
Lanzhou University
Lanzhou, Gansu
China

Shen Yonghe
Bureau of Geology and Mineral
Resources of Shanxi
Taiyuan, Shanxi
China

Sheng Jinzhang
Nanjing Institute of Geology and
Palaeontology, Academia Sinica
Nanjing
China

Su Pu
Dept. of Geology,
Shanxi Mining College
Taiyuan, Shanxi
China

Tan Zhengxiu
Hunan Regional Geological Survey Team
Jiangshe, Xiangtan
Hunan
China

Tang Wensong
Research Institute of Petroleum
Exploration and Development
P.O. Box 910, Beijing
China

Tang Yi
Nanjing Institute of Geology and
Mineral Resources
534 Zhongshan East Road, Nanjing
China

Tang Zhongli
Bureau of Geology and Mineral Resources
of Gansu Province
Dingxi Central Road, Lanzhou
Gansu
China

Tian Baolin
Beijing Graduate School,
China Institute of Mining and
Technology
Beijing
China

Tian Shuhua
562 Comprehensive Team,
Chinese Academy of Geological Sciences
Yanjiao, Sanhe
Hebei
China

Tian Zaiyi
Research Institute of Petroleum
Exploration and Development
P.O. Box 910, Beijing
China

Tong Youde

Advisory Committee of Technology
Ministry of Coal Industry
Hepingli, Beijing
China

Tong Yuming
Changsha Institute of Geotectonics,
Academia Sinica
Tongzipo, Changsha
Hunan
China

Tong Zhengxiang
Chengdu Institute of Geology and
Mineral Resources
Chengdu, Sichuan
China

Tu Guangchi
Institute of Geochemistry,
Academia Sinica
Guiyang, Guizhou
China

Wan Chaoyuan
Bureau of Geology and Mineral Resources
of Guizhou
Beijing Road, Guiyang
Guizhou
China

Wang Chengyuan
Nanjing Institute of Geology and
Palaeontology, Academia Sinica
Nanjing
China

Wang Deyou
Henan Institute of Geological Sciences
28 Jinshui Road, Zhengzhou
Henan
China

Wang Hongdi
Guizhou Regional Geological Survey Team
Huishui, Guizhou
China

Wang Hongzhen
Beijing Graduate School,
Wuhan College of Geology
Beijing
China

Wang Hui
Lanzhou Institute of Geology,
Academia Sinica
Lanzhou, Gansu
China

Wang Huiji
Natural Museum of Shanghai
Shanghai
China

Wang Keliang
Nanjing Institute of Geology and
Palaeontology, Academia Sinica
Nanjing
China

Wang Mingzhou
Section of Stratigraphy and Palaeontology,
Xi'an College of Geology
Yanta Road, Xi'an
Shaanxi
China

Wang Cong
Bureau of Geology,
Ministry of Coal Industry
Fanyang Road, Zhuozhou
Hebei

China

Wang Rennong
No. 1 Geological Prospecting Company
Ministry of Coal Industry
Handan, Hebei
China

Wang Xinping
Geology Dept., Beijing University
Beijing
China

Wang Shulin
Chinese Petroleum Society
P. O. Box 766, Beijing
China

Wang Zengji
Institute of Geology,
Chinese Academy of Geological Sciences
Beijing
China

Wang Zhi
Institute of Exploration and
Development,
Petroleum Administration of Xinjiang
Karamay, Xinjiang
China

Wang Zhihao
Nanjing Institute of Geology and
Palaeontology, Academia Sinica
Nanjing
China

Wang Zhongtang
Geological Bureau,
Ministry of Coal Industry
Zhuozhou, Hebei
China

Wei Binxian
Dept. of Geology and Mineral
Resources, Ministry of Geology
and Mineral Resources
Xisi, Beijing
China

Wei Jiayong
Guizhou Regional Geological
Survey Team
Huishui, Guizhou
China

Wei Lingdun
Guangxi Institute of Geology
1 Jianzheng Road, Nanning
Guangxi
China

Wei Weilie
Guilin College of Geology
Pingfengshan, Guilin
Guangxi
China

Wu Haoruo
Institute of Geology,
Academia Sinica
Beijing
China

Wu Ren
No. 155 Team,
Bureau of Coal Geology of
Northeast China
Shenyang, Liaoning
China

Wu Qi
Fujian Regional Geological Survey Team
Gongnong Road, Sanming

Fujian
China

Wu Wangshi
Nanjing Institute of Geology and
Palaeontology, Academia Sinica
Nanjing
China

Wu Xiantao
Jiaozuo Mining College
Jiaozuo, Henan
China

Wu Xiuyuan
Nanjing Institute of Geology and
Palaeontology, Academia Sinica
Nanjing
China

Wu Yaocheng
Dept. of Geology,
Nanjing University
Nanjing
China

Xia Guoying
Tianjin Institute of Geology and
Mineral Resources
26 Jintang Road, Tianjin
China

Xiang Liwen
Institute of Geology,
Chinese Academy of Geological Sciences
Beijing
China

Xiao Weimin
Guizhou Regional Geological Survey
Team
Huishui, Guizhou
China

Xu Daoyi
Institute of Geology,
State Seismological Bureau
Beijing
China

Xu Huilong
Shanxi Company of Coalfield
Geology and Exploration
Wucheng Road, Taiyuan
Shanxi
China

Xu Jiamo
No. 1 Geological Corporation,
Ministry of Coal Industry
Handan, Hebei
China

Xu Tongchun
Institute of Geophysics,
State Seismological Bureau
Beijing
China

Yang Fengqing
Dept. of Geology,
Wuhan College of Geology
Wuchang, Hubei
China

Yang Guanxiu
Beijing Graduate School,
Wuhan College of Geology
Beijing
China

Yang Jiduan
Institute of Geology,
Chinese Academy of Geological Sciences

Beijing
China

Yang Jiawen
No. 3 Geological Team of Yunnan
Xiangshan East Road, Dali
Yunnan
China

Yang Jingzhi
Nanjing Institute of Geology and
Palaeontology, Academia Sinica
Nanjing
China

Yang Meixia
Bureau of Coal Geology of Northeast
China
Shenyang, Liaoning
China

Yang Qi
Beijing Graduate School,
Wuhan College of Geology
Beijing
China

Yang Shipu
Beijing Graduate School,
Wuhan College of Geology
Beijing
China

Yang Xilu
Institute of Geology and Exploration,
CCMRI, Ministry of Coal Industry
Xi'an, Shaanxi
China

Yang Zunyi
Beijing Graduate School,
Wuhan College of Geology

Beijing
China

Yao Yimin
Geological Research Institute of
Shengli Oilfield
Dongying, Shandong
China

Ye Sujuan
Institute of Geology,
Academia Sinica
Beijing
China

Yin Bao'an
Guangxi Regional Geological Survey
Team
Yishan, Guangxi
China

Yin Jixiang
Institute of Geology,
Academia Sinica
Beijing
China

Yu Xiurong
Geological Publishing House
Beijing
China

Yuan Yaoting
Institute of Geology and
Exploration,
CCMRI, Ministry of Coal Industry
Xi'an, Shaanxi
China

Yue Xixin
Ministry of Geology and Mineral
Resources

Xisi, Beijing
China

Zeng Daofu
Southwest Petroleum Geological
Bureau
Qinlongchang, Chengdu
Sichuan
China

Zeng Dingqian
Western South Chian Sea Petroleum
Company
P. O. Box 11, Puotou, Zhanjiang
Guangdong
China

Zhan Lipei
Institute of Geology,
Chinese Academy of Geological Sciences
Beijing
China

Zhang Aiyun
Beijing Graduate School,
Wuhan College of Gology
Beijing
China

Zhang Dajiang
Research Institute of Petroleum
Exploration and Development
P. O. Box 910, Beijing
China

Zhang Chuangan
Research Institute of Petroleum
Exploration and Development
P. O. Box 910, Beijing
China

Zhang Linxin
Nanjing Institute of Geology and
Palaeontology, Academia Sinica
Nanjing
China

Zhang Pengfei
Beijing Graduate School,
Chinese Institute of Mining and
Technology
Beijing
China

Zhang Renjie
Yichang Institute of Geology and
Mineral Resources
P. O. Box 502, Yichang
Hubei
China

Zhang Rumei
Science Press
Beijing
China

Zhang Shanzhen
Nanjing Institute of Geology and
Palaeontology, Academia Sinica
Nanjing
China

Zhang Shouxin
Institute of Geology,
Academia Sinica
Beijing
China

Zhang Shuxiu
Institute of Geology,
Chinese Academy of Geological Sciences
Beijing
China

Zhang Tielin
Bureau of Geology and Mineral
Resources of Shanxi
5 Bingzhou Road, Taiyuan
Shanxi
China

Zhang Zhenghua
Guizhou Petroleum Prospecting
Ministry of Petroleum Industry
1 Qianling East Road, Guiyang
Guizhou
China

Zhang Zhifei
National Natural Science
Foundation of China
Beijing
China

Zhang Zhimin
Institute of Geology Science,
Bureau of Geology and Mineral
Resources of Xinjiang
Youhao Road, Urumqi
Xinjiang
China

Zhang Zuqi
Dept. of Geology,
South-Central Polytechnic University
Yuelushan, Changsha
Hunan
China

Zhao Guoquan
Tiangjin Institute of Geology,
Ministry of Metallurgical Industry
Tianjin
China

Zhao Longyie
Beijing Graduate School,
Wuhan College of Geology
Chengfu Road, Beijing
China

Zhao Ruxuan
Institute of Geological Sciences of
Guangdong
739 Dongfeng East Road, Guangzhou
Guangdong
China

Zhao Shiqing
Dept. of Geology,
Huainan Mining Institute
Huainan, Anhui
China

Zhao Xiuhu
Nanjing Institute of Geology and
Palaeontology, Academia Sinica
Nanjing
China

Zhao Zhixin
Research Institute of Petroleum
Exploration and Development,
Xinjiang Petroleum Administration
Karamay, Xinjiang
China

Zheng Hong
Beijing Graduate School,
Wuhan College of Geology
Beijing
China

Zheng Luren
National Committee of Reserves of
Mineral Resources
Xisi, Beijing
China

Zhong Rong
Institute of Geomechanics,
Chinese Academy of Geological Sciences
Qinghuasi, Beijing
China

Zhong Xianliang
Dept. of Geology,
Fuxin Mining College
Fuxin, Liaoning
China

Zhong Xiaochun
Research Institute of Petroleum
Exploration and Development
P. O. Box 910, Beijing
China

Zhou Mingjian
Science Press
Beijing
China

Zhou Shiyong
Dept. of Chemical Industry,
Anshan Institute of Steel and
Iron Technology
Anshan, Liaoning
China

Zhou Xinhu
Zhejiang Institute of Petroleum
and Geology
Hangzhou, Zhejiang
China

Zhou Yiping
Kunming Institute of Coal Science
21 Minhang Road, Kunming
Yunnan
China

Zhou Zhongyi
Institute of Geochemistry,
Academia Sinica
P. O. Box 91, Guiyang
Guizhou
China

Zhou Zuren
Nanjing Institute of Geology and
Palaeontology, Academia Sinica
Nanjing
China

Zhu Deyuan
Comprehensive Institute of Petroleum
and Geology, Ministry of Geology and
Mineral Resources
P. O. Box 12, Jiangling
Hubei
China

Zhu Hongyuan
Dept. of Geology,
Jianghan Oil College
Jiangling, Hubei
China

Zhuang Jun
Institute of Geology and Exploration,
CCMRI, Ministry of Coal Industry
Xi'an, Shaanxi
China

ARGENTINA (1)

Archangchsky, Sergio
1132 Urquiza Vicente Lopez,
Buenos Aires, Argentina 1638

AUSTRALIA (7)

Brakel, A. T.
Bureau of Mineral Resources
GPO Box 378, Canberra ACT,
Australia 2601

Cook, A. C.
University of Wollongong
Northfields Ave.
Wollongong, N.S.W. 2500
Australia

Dickins, James MacGregor
Bureau of Mineral Resources
P. O. Box 378, Canberra ACT, 2601
Australia

Engel, Brian A.
Dept. of Geol.,
Univ. of Newcastle
New South Wales, 2308
Australia

Rigby, John
Dept. of Mines, Queensland,
Minerals and Energy Centre
P. O. Box 194
Brisbane 4001 Queensland
Australia

Webb, Gregary E.
Dept. Geology, Univ. Queensland
St. Lucia, Queensland 4067
Australia

Ward, Colin R.
Dept. of Applied Geol.
School of Mines
The Univ. of New South Wales
P. O. Box 1, Kensington
New South Wales, 2033
Australia

BELGIUM (3)

Bouckaert, Jozef
Belgian Geological Survey
13 Jenner, Brussels, 1040
Belgium

Pierart, Pierre
Service de Biologie
Mons University, 24
Avenue du Champ de Mars
7000 Mons, Belgium

Somers, Yvonne A. M.
Iniex 200 rue du Chera
Liege, Belgium B-4000

BRITAIN (9)

Barclay, William J.
British Geological Survey
Keyworth, Nottingham, NG12 5GG
England, UK

Browne, M. A. E.
The British Geological Survey
Murchison House, West Mains Road
Edinbugh EH9 3LA
Scotland, UK

Higgins, A. C.
The British Petroleum Corp.
BP Research Centre, Chertsey
Road, Sunbury-on-Thames
Middlesex TW16 7LN
England, UK

Owens, Bernard
British Geological Survey

Keyworth, Nottingham, NG12 5GG
England, UK

Ramsbottom, W. H. C.
Dept. Geology,
The Univ. of Sheffield, Mappin Street
Sheffield S1 3JD
England, UK

Scott, A. C.
Geology Dept. Royal Holloway,
Bedford, New College
Univ. of London
Egham, Surrey, TW200EX
England, UK

Smith, H. W.
British Coal, 16 Silverdale,
Close Sheffield
South Yorkshire, S11 9IN
England, UK

Tarling, D. H.
School of Physics,
The Univ. Newcastle-upon-Tyne
NE1 7RU, UK

Varker, W. J.
Dept. of Earth Sci.,
The Univ. Leeds LS2 95T
England, UK

CANADA (12)

Bustin, R. M.
Dept. of Geological Sciences,
Univ. British Columbia
Vancouver, B. C. V6T 2B4
Canada

Danner, W. R.
Dept. of Geological Sciences,
Univ. British Columbia
6339 Stores Road, Vancouver,
V6T 2B4
British Columbia, Canada

Grant, R. H.
Dept. Geology,
Univ. of New Brunswick
P. O. Box 4400, Fredericton
N. B. E3B 5A3, Canada

Henerson, C. M.
Dept. Geology and Geophysics,
Univ. Calgary, Calgary
Alberta, T2N 1N4, Canada

Lapointe, Martine
1945 Decelles, Saint-Laurent
Quebec, Canada H4M 1B1

Mamet, Bernard Leon
Dept. Geol.,
Univ. Montreal Nontreal, P. O. C. P. 6128
Canada H3C 3J7

Orchard, Michael J.
Geological Survey Canada
100 West Pender
Vancouver B. C., Canada V6B 1R8

Pickerill, R. K.
Dept. of Geology,
Univ. of Rovunswick
P. O. Box 4400, Fredericton
N. E. E3B 1K7, Canada

Richards, Barry C.
Institute of Sedimentary
and Petroleum Geology
Geological Survey of Canada
3303-33rd Str., N. W. Calgary
Alberta, Canada T2L 2A7

Ross, J. V.
Dept. Geological Sciences,
Univ. British Columbia
Vancouver, British Columbia,
V6T 2B4, Canada

Utting, John
Institute of Sedimentary
and Petroleum Geology
Geological Survey of Canada
3303-33rd Str. N. W. Calgary
Alberta, Canada T2L 2A7

Zodrow, E. L.
Univ. of College of Cape,
Breton, Dept. Geology, Glage Bay,
Highway Sydney
Nova Scotia B1P 6L2 Canada

CZECHOSLOVAKIA (3)

Miroslav, K.
Fuel Research Institute
Prague 9-Bechovice, 25079
Czechoslovakia

Vachtl, Bohumil
Plaminkove 1581 Praha 4
Czechoslovakia 14000

Vozar, J.
Slovensky Geologicky urad
Foreign Relation Dept.
Bukurestska 4.
Bratislava 817 62
Czechoslovakia

EGYPT (1)

Kholief, M. M.
Exploration Dept.
Petroleum Research Institute
Nasr City 7th Region, Cario
Egypt

FRANCE (5)

Alpern, B.
23 bis rue des Cordelieres
Paris 75013, France

Beaudoin, Bernard
Ecole des Mines de Paris
Laboratoire de Sedimentologie
35 rue Saint-Honore, 77305
France

Laveine, Jean-Pierre
Univ. des Sciences et
Techniques, de Lille UFR
Sciences de la Terre
Villeneuve D'Ascq 59655
France

Legrand-Blain, Marie
Tauzia Rtede
Bayonne-33170-Gradignan
France

Tien, Nguyen Duc
Total C.F.P.
92069 Paris La Defense
Cedex 47, France

F. R. GERMANY (28)

Altermann, W.
Frege Univ. of Berlin
Institute of Applied Geology
Wichernstr. 16, D-1000
Berlin 33, FRG

Brauckmann, Carsten Gunter

Fuhlrott-Museum, 20 Auer
Schulstrasse
D-5600 Wuppertal 1, FRG

Dahm, H. D.
Geologisches Landesamt NW
De-Greiff-Str. 195
Postfach 1080, D-4150 Krefeld
FRG

Engel, H.
Herzstr 32, 6688 Illingen 2
FRG

Hansel, G.
Ruhrkohle Ag, Rellinghauser,
Str. 1, 4300, Essen 1, FRG

Hedemann, H. A.
Geologisches Institut der
Univ., Schlossgarten 5, D-8520
Erlangen FRG

Herbig, H. G.
Instiut fuer Geologie
Freie Univ. Berlin
Altensteinstrasse 34A D-1000
Berlin 33, FRG

Hiltmann, W.
Budnesanstalt fuer
Geowissenschaften und Rohstoffe
D-3000 Hannover 51, FRG

Horn, M.
Hessisches Landesamt fuer
Bodenforschung, Leberberg 9
D-6200 Wiesbaden, FRG

Josten, K. H.
Gelolgisches Landesamt
Norfreim Westfalen
415 Krefeld, Postfach 1080

De-Greiff-Str. 195
FRG

Juch, D. D.
Geologisches Landesamt NW
De-Greiff-Str. 195
D-4150 Krefeld, FRG

Klitzsch, E. H.
TU Berlin, Sekr. SFB 69
Ackerstrasse 71-76, D-1000
Berlin 65, FRG

Kullmann, Jurgen
Geol. Palaont. Institut
Sigwartstr. 10, D-7400
Tubingen, FRG

Kutzner, R.
Ruhrkohle AG, Habichtstr 60
Essen 1, D-4300, FRG

Mosbrugger, V. J.
Palaeontologisches Institut
Nussallee 8, D-5300 Bonn, FRG

Niemoller, B.
Stephanstrasse, 8
D-4132 Kamp-Lintfart, FRG

Paproth, E. E.
Geologiches Landesamt NW
De-Greiff-Str. 195
Krefeld D-4150, FRG

Plein, E. P. O.
Dudweiler Str. 3
Hannover 71, D-3000, FRG

Prashnowsty, A. A.

Geologisches Institut
Pleicherwall 1, 8700 Wuerzburg
FRG

Rabitz, A.
Geologisches Landesamt NW
De-Greiff-Str. 195 Krefeld,
D-4150, FRG

Reiche, E.
Geologisches Landesamt NW
De-Greiff-Str. 195
D-4150, Krefeld, FRG

Riegel, W.
Institut and Museum fuer
Geologie und Palaeotologie
Goldschmidtstr 3,
3400 Gottingen, FRG

Sauer, A. F.
c/o Bergbau ag Westfalen
Silberstr 22, P. O. Box 872
D-4600 Dortmund 1, FRG

Schweitzer, H. J.
Institut fuer Palaeontologie
8 Nussallee, 5300 Bonn, FRG

Steeler, M.
Wachalderbusch 11 Velbert
5620, FRG

Von Sperber, M.
Westf Berggewerkschaftskasse (IAG)
Herner Str. 45, D-4630
Bochum, FRG

Wolf, M.
Lehrstuhl fur Erdol und Kohle
RWTH Aachen Lochnerstrasse,
4-20, D-5100, Koln 1, FRG

Wopffner, H.
Geologisches Instiut
Univ. zu Koln, Zulpicherstrasse 49
D-5000, Koln 1, FRG

D. R. GERMANY (2)

Franke, D.
Central Geological Institute
44 Invalidenstrasse
Berlin 1040, GDR

Hoffmann, N.
Central Geological Institute
44 Invalidenstrasse
Berlin 1040, GDR

INDIA (1)

Chandra, Debabrata
Indian School of Mines
Dhanbad Bihar, India 826004

IRELAND (1)

Sevastopulo, G. D.
Geology Dept., Trinity College
Dublin, Ireland

ISRAEL (2)

Gvirtzma, G.
Geological Survey of Israel
30 Malkhe Yisrael St.
Jerusalem 95501, Israel

Kafri, U.
Geological Survey of Israel
30 Malkhe Yisrael St.
Jerusalem 95501, Israel

ITALY (1)

Grignani, Dario
AGIP S. P. A. /STIG.
P. O. Box 12069-20120 Milano
Italy

JAPAN (16)

Adachi, Shuko
Institute of Geoscience,
The Univ. of Tsukuba
Ibaraki 305, Japan

Aihara, Atsuo
Dept. of Geology, Fuchlty of Science,
Kyushu Univ.
Hakozaki 6-10-1
Higashiku, Fukuoka, Japan 812

Ezaki, Y.
Dept. Geology and Mineralogy
Faculty of Science, Hokkaide Univ.
N10 W8 Kita-Ku
Sapporo 060, Japan

Hisada, Ken-ichiro
Dept. of Geology. Osako Kyoiku Univ.
Tennohji-ku, Osaka 543
Japan

Igo, Hisaharu
Dept. of Earth Sciences,
Tokyo Gakugei Univ.
Koganei 184, Tokyo, Japan

Igo, Hisayoshi
Institute of Geoscience,
The Univ. Tsukuba
Ibaraki 305, Japan

Kato, M.
Dept. of Geology and Mineralogy,
Hokkaido Univ., Faculty of Science
Sapporo, Japan

Kimura, T.
Tokyo Gakugei Univ.
Koganei, Tokyo, Japan 184

Koike, T.
Geological Instute,
Yokohama National Univ.
156 Tokiwadai, Hodogaya-ku
Yokohama 240, Japan [(45) 335-1451]

Nakamura, K.
Dept. of Geology and Mineralogy,
Hokkaido Univ., Faculty of Science
Sapporo 060 Japan

Nakazawa, Keiji
Faculty of General Education,
Kinki Univ.
Kowakae, 3-4-1
Higashiosaka, Japan 577

Okimura, Y.
Institute of Geology and
Mineralogy, Faculty of Science
Hiroshima Univ.
1-189 Higashi Senda
Hiroshima, 730, Japan

Ota, M.
Kitakyushu Museum of Natural History,
Yahata Station Bldg. 6-1,
Nishihonmachi 3-Chome
Yahatahigashiku
Kitakyushu, 805, Japan

Ota, Y.
Dept. of Geology, Faculty of Science,

Kyushe Univ.
6-10-1 Hakozaki-cho, Higashi-ku
Fukuoka 812, Japan [(92) 641-1101]

Sugiyama, Tetsuo
Dept. of Geology, Faculty of Science,
Fukuoka Univ.
8-19-1 Nanakuma Ihonan-ku, Fukuoka
Japan 814-01

Yanagida, J.
Dept. of Geology,
Faculty of Science, Kyushu Univ.
33 6-10-1 Higashi-ku, Fukuoka
812 Japan

D. P. R. KOREA (3)

Kim Jong Heui
The Geological Research,
Institute of Academy Science
D. P. R. K.

Em Hye Yong
The Geological Research,
Institute of Academy Science
D. P. R. K.

Yun Bong Jun
The Geological Reserach,
Institute of Academy Science,
D. P. R. K.

R. O. KOREA (1)

Chun, H. Y.
Korea Institute of Energy
and Resources
219-5 Garibong-dong Guro-gu,
Seoul, 150-60,Korea

MADAGASCAR (1)

Rakotoarivelo, Heri J.
Faculte des Sciences
Serice Paleontologie
Boite Postale 906
101 Antananarivo
Madagascar

MALAYSIA (1)

Metcalfe, Ian
Dept. of Geology, Univ. Kebangsaan,
Malaysia
Bangi, Selangor, Malaysia

MOZAMBIQUE (1)

Vasconcelos, L.
Dept. Geologia-UEN
C. P. 257-Maputo, Mozambique

THE NETHERLANDS (4)

Conijn, Ijsbrand F.
Oranjelaan 4, Melick
Holland 6074 AV

Krans, Th. F.
Geological Bureau,
Voskuilenweg 131
6416 AJ Heerlen, The Netherland

Van Tongeren, P. C. H.
Geological Bureau,
Voskuilenweg 131
6416 AJ Heerlen, The Netherland

Winkler Prins, Cor F.
Rijksmuseum var Geologie en
Mineralogie

Hooglandse kerkgracht 17
Leiden, Holland 2312 HS

NEW ZEALAND (1)

Black, Philippa M.
Dept. of Geology, Univ. Auckland
Private Bog Auckland
New Zealand

POLAND (9)

Dybova-Jachowicz, S.
Instytut Geologiczny
Rakowiecka 4, 00-975 Warszawa
Poland

Jachowicz, Alesander
Univ. Slaski, ul. Bankowa 12
40-007 Katowice
Poland

Jaworowski, Krzysztof
Instytut Geologiczny
Rakowiecka 4, 00-975 Warszawa
Poland

Kozlowski, Azimierz
Univ. Slaski, ul. Bankowa 12
40-007 Katowice
Poland

Musiat, L.
Instytut Geologiczny
Rakowiecka 4, 00-975 Warszaw
Poland

Peryt, Tadevsz
Instytut Geologiczny
Rakowiecka 4, 00-975 Warszawa
Poland

Porzycki, Jozff
Geological Institute, ul. Bialego 1
41-200 Sosnowiec
Poland

Zelichowski, Antoni M.
Instytut Geologiczny
Rakowiecka 4, 00-975 Warszawa
Poland

Zuberek, Waclaw
Univ. Slaski, ul. Bankowa 12
40-007 Katowice
Poland

PORTUGAL (1)

Oliverira, Jose Tomas
Servicos Geologicos de
Portugal
rua da Academia das Ciencias
19, 2 1200 Lisboa-Portugal

SPAIN (5)

Alvarez-Ramis, Concepcion
Laboratorio de Paleobotanica
Dept. de Paleontologia Facultad
de Ciencias
Geologicas, U. C. M. Madrid
Spain 28040

Granados, L. F. G.

Garcia-Loygorri, A.
Instituto Geologico y Minero,
de Espana, Rio Rosas
23 Madrid-3, Espana

Virgili, Carmina
Colegio de Espanā
Cité Internatonale Universitaire de Paris

7 Boulevard Jourdan
75041 Paris

Wagner, R. H.
Lista de Correos Cervera de
Pisuerga (Palencia)
Spain

SWEDEN (1)

Bengtsson, Margaretha E.
M. Bengtsson Consultant AB
Blacksvampvagen 16, S-611 63
Nykoping, Sweden

SWITZERLAND (2)

Buser, Hugo
Geological Am.,
Unterterzen, CH8882
Switzerland

Remane, J. R.
Univ de Neuchatel
Institut de Geologie
11 rue Emile-Argand, CH-200
Neuchatel 7
Switzerland

U. S. A. (52)

Anderson, J., Jr.
Dept. of Geology and
Planetany Science
Univ. of Pittsburgh
Pittsburgh, PA 15260, USA

Baxter, S.
Ohio University at Lancaster
Lancaster, Ohio 43130, USA

Beus, S. S.
Dept. of Geology, P. O. Box 6030
Northern Arizona Univ.
Flagstaff, AZ 86011, USA

Brenckle, Paul L.
Amoco Production Company
Research Centre
P. O. Box 3385 Tulsa
Olkahoma, USA 74102

Chesnut, Donald R., Jr.
Kentucky Geological Survey
Univ. Kentucky
311 Breckinridge Hall
Lexington, Kentucky 40506-0056
USA

Chou Chen-lin
Illinois State Geological Survey
615 East Peabody Drive
Champaign Illinois 61820, USA

Cocke, Julius M.
Central Missouri State Univ.
Geol. Dept., C. M. S. U.
Warrensburg, Missouri
USA 64019

Conkin, J. E.
Dept. of Geology
College of Arts and Science
Univ. of Louisville
Belknap Campus, Louisville
Kentucky 40292, USA

Davis, A.
Coal Research Section
The Pennsylvania State Univ.
513 Deike Building
Univ. Park, PA 16801, USA

Dewey, C. P.

Dept. of Geology and Geography
P. O. Drawer 5167
Mississippi State Univ.
Mississippi 39762, USA

Donaldson, Alan
Geology and geography
Department,
West Virgini University
Morgantown, W. V., USA

Dutro, J. Thomas, Jr.
Room E-316
Museum of Natural History
Washington, D. C. 20560, USA

Englund, Kenneth J.
U. S. Geological Survey
Reston, VA 22092, USA

Ettonsohn, Frank John
Geol. Dept., Univ. of Kentucky
Lexington, Kentucky 40506, USA

Gastaldo, R. A.
Dept. of Geology,
Auburn Univ.
210 Petrie Hall
Alabama 36849-3501, USA

Gilmour, Ernest H.
Dept. of Geology,
Eastern Washington Univ.
Cheney, Washington 99004, USA

Gordon, M., Jr.
U. S. Geological Survey
E-501, U.S. National Museum
Washington, D. C. 20560, USA

Grant, Richard E.
Smithsonian Institution, E-206
Natural History Building
Washington, D. C. 20560, USA

Hasson, K.
Dept. of Geography and Geology,
East Tennessee State Univ.
College of Arts and Sciences
Box 22870A
Johnson City, Tennessee, USA

Heckel, P. H.
Dept. of Geology,
The Univ. of Iowa
Iowa City, Iowa 52242, USA

Johnson, K. S.
Oklahoma Geological Survey
830 van Vleet Oval
RM 163/Norman, Oklahoma 73019
USA

Jorden, William M.
Millersville Univ.
PA 17551, USA

Keiser, Alan F.
West Virginia Geological
and Economic Survey
P. O. Box 879, Morgantown
WV 26507-0879, USA

Lane, R.
Amoco Production Company
Research Center
P. O. Box 3385, Tulsa
OK 74102, USA

Langenheim, R. L., Jr.
Dept. of Geology,
Univ. of Illinois
245 NHB, Illinois 61801, USA

Leary R. L.
Illinois State Nuseum
Spring and Edwards
Springfield
Illinois 62706, USA

Lim, S. W. M.
Utah International Inc.
510-C Herdon Parkway
Virginia 22070, USA

Lockley, Martin G.
Dept. Geol.
Univ. Colorado at Denver
1100 14th Street Denver
Colorado 80202, USA

Lyons, Paul C.
U. S. Geological Survey
956 National Center
Reston, VA 22092, USA

Manger, W. L.
Dept. of Geology,
Univ. of Arkansas
Fayetteville
Arkansas 72701, USA

Maples, Christopher G.
Dept. of Earth Science
College of Sciences
Clemson Univ., Brackett Hall
Clemson, South Carolina
29634-1908, USA

McCollum, L.
Geology Dept., Eastern Wa. U.
Cheney, WA 99004, USA

Merriam, Daniel F.
Dept. of Geology,
Wichita State Univ.
Wichita, Kansas 67208, USA

Milligan, E. N.
Utah International Inc.
510-C Herndon Parkway
Virginia 22070, USA

Neal, Dorald W.
Dept. of Geology,
East Carolina Univ.
Greenville
North Carolina 27834, USA

Nestell, Merlynd K.
Dept. of Geology
Univ. Texas at Arlington
Arlington, Texas 76019, USA

Renton, John
1424 Dogwood Ave. J.
Morgantown, W. V., 26505, USA

Ross, Charles A.
Chevron, USA
Inc. P. O. Box 1635
Housten, Texas 77251, USA

Ross, June R. P.
Dept. of Biology,
Western Washington Univ.
Bellingham, Washington 98225
USA

Sando, W. J.
E-501, US National Museum
of Natural History
Washington, D. C. 20560, USA

Schram, Frederick R.
San Viego,
Natural History Museum
P. O. Box 1390, San Diego

CA 92112, USA

Skipp, B.
U. S. Geological Survey
Box 25064 M. S. 913
Denver Federal Center
Denver, CO 80225, USA

Snyder, Walter S.
Dept. of Geology and Geophysics,
Biose State Univ.
1910 University Drive,
Boise, Idaho 83725, USA

Spinosa, Claude
Dept. of Geology and Geophysics,
Boise State Univ.
1910 University Drive,
Boise, Idaho 83725, USA

Sutherland, P. K.
School of Geology. and Geophysics,
Univ. of Oklahoma, Norman
Oklahoma 73019, USA

Tidwell, W. D.
Dept. of Botany, Brihgam Young Univ.
401 Widb, Provo, Utah 84602, USA

Waters, Johnny A.
Dept. of Geology,
West Georgia College
Carrollton, GA 30118, USA

Webb, Fred, Jr.
Dept. of Geology,
Appalachian State Univ.
Boone, North Carolina
28608, USA

Webster, B. E.
Dept. of Geology,
Washington State Univ.
Pullman Washington 99164-2812
USA

Weibel, C. P.
Illinois State Geological Survey,
Univ. Illinois 615,
East Peabody Drive
Champaign, Illinois 61820, USA

Windolph, John
U. S. Geological Survey
Reston, VA 22092, USA

Yochelson, E. L.
E-501 National Museum
of Natural History
Washington, D. C. 20560, USA

U.S.S.R. (4)

Cherepovski, V. F.
Ministry of Geology
Moscow, USSR

Chuvashov, Boris 1.
Institute of Geology
and Geochemistry
of Urals Section, Ac. Sci. USSR
620219 Sverdlovsk
Pochtovyi per. 7
USSR

Kotljar, G. V.
ul. Mozajacaya, 40-4
198147 Laningrod, USSR

Simakov, K. V.
N. E. Interdiscipl. Sci. Res. Inst.,
Far East Sci. Center
U. S. S. R. Academy of Sciences
Portovajy ul. 16
685005 Magadan, USSR

XIe Congrès International de Stratigraphie et de Géologie du Carbonifère Beijing 1987,
Compte Rendu 1 (1991): 75-77

Non-attending Members

Bouroz, Alexis
110 Av., Felix Faure
75015 Paris
France

British Geological Survey
Library Book Ordering
Keyworth
Nottingham NG12 5GG
England

The British Library
Sci. Ref. Libr.
25 Southampton Buildings
Chancery Lane
London, WC2A 1AW
UK

British Museum Natural History
No 2 A/C, Library Services
Cromwell Rd., London
SW7 5BD, UK

Bureau de Recherches
Geologiques et Minieres
Section DOC/Achats
B.P. 6009, 45060 Oreeans Cedex
France

Chang, Li-sho
3406 Redbud Lane
Raleigh, N.C. 27607
USA

Dipartmento di Scienza Della Terra
Universita degli studi di siena
via delle cerchia 3
53100 SIENA SI
Italy

Garcia, Sergio Rodriguez
Depatamento de paleontologia
facultad de ciencias geologicas
28040 Madrid
Spain

Geological and Paleontological
Institute of the Phillipps
University at Marburg/Lahn
FRG

Geological Society
Burlington House
London, W1V OJU
UK

Geology Library
Univ. of Texas
Austin, Texas
USA

Green Library
Stanford University
Stanford, CA 94305
USA

Institut de Geologie
Universite de Liege
7, place du 20 Aout
4000, Liege

Belgique

Institute of Sedimentary
and Petroleum Geology
3303-33rd Street N.W.
Calgary, Alberta
Canada T2L 2A7

Institut und Museum fuer Geologie
und Palaeontologie
Univ. Gottingen
FRG

Kuwano, Y.
Dept. of Palaeontology,
National Sci. Museum
3-23-1, Hyakunin-cho
Shinjuku, Tokyo 160
Japan

Lemos de Sousa, M.J.
Mineralogia e Geologia
Faculdade de Ciencas
Praca de Gomes Teixeira
4000 Porto--Portugal

Ramovs, A.
Katedra za geologijo in
paleontologijo, Askerceva
YU-61000, Ljubljana
Yugoslavia

Royal Holloway and Bedford
New College
Univ. of London
Egham Hill, Egham
Surrey TW20 OEX
UK

Sada, K.
Faculty of Integrated
Arts and Sciences,
Hiroshima Univ.
Hiroshima 730
Japan

Sakagami, Sumio
Chiba Univ.
Dept. of Earth Sciences
Faculty of Science, Yayoi-cho
CHIBA, 280
Japan

Sartenaee, P.
Institut Royal des Sci.
Naturelles de Belgique
rue Vautier 29
B-1040, Bruxelles
Belgium

Senckenbergische Naturforschende
Gesellshaft, Senckenberganlage 25
D-6000 Frankfurt 1
FRG

Taylor, Thomas N.
Dept. of Botany
The Ohio State University
1735 Neil Av.
Columbus, Ohio 43210-1293
USA

Taylor, Edith L.
Dept. of Botany
The Ohio State University
1735 Neil Av.
Columbus, Ohio 43210-1293
USA

Trinity College
The Univ. of Dublin
Ireland

Univ. of Kentucky Libr.

Margaret, I King Libr.
Lexington KY 40506
USA

Univ. of Queensland
St. Lucia, QLD 4067
Australia

Vai, G. Battista
Istituto di Geologie e
Paleontologia, Via zamboni
67-40127 Bologna
Italy

Waterhouse, John B.
Dept. of Geology,
Univ. of Queensland
St Lucia, Brisbane
Australia 4067

Watson Library
Univ. of Kansas Libr.
Lawrence, KS 66045-2800
USA

XIe Congrès International de Stratigraphie et de Géologie du Carbonifère Beijing 1987,
Compte Rendu 1 (1991): 78-79

Congress Sections

1. Stratigraphy and Geochronology

Chairmen: He Xilin (China), K. Nakamura (Japan), P.K. Sutherland (USA), Hisayoshi Igo (Japan)
Coordinator: Guo Shengzhe, Liu Fa

2. Palaeontology, Palaeoecology and Palaeobiogeography

2a. Palaeozoology

Chairmen: J.T. Dutro (USA), M. Kato (Japan), W.J. Sando (USA)
Coordinator: Zhang Renjie

2b. Palaeobotany

Chairmen: J. Rigby (Australia), S. Archangelsky (Argentina), Li Xingxue (China), H. J. Schweitzer (FRG), J.P. Laveine (France)
Coordinator: Shen Guanglong, Duan Shuying

2c. Micropalaeontology

Chairmen: B.L. Mamet (Canada), A.C. Higgins (UK), P.L. Brenckle (USA)
Coordinator: Wang Chengyuan, Zheng Hong

2d. Palaeoecology and Palaeobiogeography

Chairmen: Wang Hongzhen (China), C. F. Winkler Prins (The Netherlands), C. Spinosa (USA)
Coordinator: Wu Haoruo, Wu Qi

3. Sedimentary and Geochemistry

Chairmen: K. Jaworowski (Poland), Fang Shaoxian (China), J.T. Oliveira (Portugal), J. Vozar (Czechoslvovakia)
Coordinator: Fang Shaoxian, Han Shuhe

4. Palaeogeography and Palaeoclimatology

Chairmen: W.H.C. Ramsbottom (UK), M. Gordon, Jr. (USA)
Coordinator: Mu Xinan

5. Economic Geology

Chairmen: E. Reiche (FRG), Th. F. Krans (The Netherlands)
Coordinator: Feng Baohua

6. Petrology and Geology of Coal

Chairmen: B. Alpern (France), A.H.V. Smith (UK), M. Wolf (FRG)
Coordinator: Jin Kuili, Chen Peng

7. Tectonics and Geophysics

Chairmen: D. Tarling (UK), R.H. Danner (Canada), Lin Jinlu (China)
Coordinator: Lin Jinlu

8. Geobiochemistry and Potential Hydrocarbon

Chairmen: Zhou Zhongyi (China), Tian Zaiyi (China)
Coordinator: Zhou Zhongyi

XIe Congrès International de Stratigraphie et de Géologie du Carbonifère Beijing 1987,
Compte Rendu 1 (1991): 80-81

Congress Symposia

1. Carboniferous-Permian Boundary

Chairmen: C.A. Ross (USA), B.I. Chuvashov (USSR), Wu Wangshi (China)
Coordinator: Rui Lin

2. Devonian-Carboniferous Boundary

Chairmen: E.E. Paproth (FRG), J. Bouckaert (Belgium)
Coordinator: Wang Chengyuan

3. Subdivision of the Carboniferous System

Chairmen: H.R. Lane (USA), W.L. Manger (USA), G.D. Sevastopule (Ireland)
Coordinator: Xu Shanhong

4. Palaeozoic Microflora

Chairmen: B.Owens (UK), J.Utting (Canada), P.Pierart (Belgium), S. Dybova-Jachowicz (Poland)
Coordinator: Ouyang Shu

5. Evolution of Tethys during Permo-Carboniferous

Chariman: R.E. Grant (USA)
Coordinator: Wu Haoruo

6. Carboniferous and Permian of Gondwanaland

Chairmen: H. Wopfner (FRG), Yin Jixiang (China)
Coordinator: Yin Jixiang

7. Exploration of Permo-Carboniferous Coal Resources

Chairman: Han Dexin (China)
Coordinator: Mao Bangzhuo

8. Global Events and Evolution during Permo-Carboniferous Time

Chairman: Yang Zunyi (China)
Coordinator: Yang Zunyi

9. Coal and Coal-forming Environment in Earth History

Chairmen: Chou Chenlin (USA), K.H. Josten (FRG), Yang Qi (China)
Coordinator: Guan Shiqiao

10. Correlation of Permian Sequences in Tethys

11. Subdivision of the Permian System

Chairmen: J.M. Dickins (Australia), K. Nakazawa (Japan)
Coordinator: Jin Yugan

XIe Congrès International de Stratigraphie et de Géologie du Carbonifère Beijing 1987,
Compte Rendu 1 (1991): 82-83

Business Meetings and Other Concurrent Symposia

Meeting of the International Permanent Committee for ICC

Chairman: Yang Jingzhi

40th Annual Meeting of ICCP

Chairman: B. Alpern
Secretary: M. Wolf

Meeting of the International Commission on Palaeozoic Microflora

Chairman: B. Owens

Meeting of the International Subcommission on Carboniferous Stratigraphy

Chairman: W. H. C. Ramsbottom

Meeting of the International Subcommission on Permian Stratigraphy

Chairmen: Sheng Jinzhang, J. M. Dickins

Meeting of the Working Group on Devonian-Carboniferous Boundary

Chairman: E. Paproth

Meeting of the Working Group on Mid-Carboniferous Boundary

Chairman: R. Lane

Meeting of the Working Group on Carboniferous-Permian Boundary

Chairman: Wu Wangshi

Meeting of the International Working Committee on Upper Permian Stratigraphy

Chairman: J.M. Dickins

Colloquium Approach a Global Stratigraphic Scheme

Chairman: Jin Yugan

XIe Congrès International de Stratigraphie et de Géologie du Carbonifère Beijing 1987, Compte Rendu 1 (1991): 84-96

THE CARBONIFEROUS SYSTEM OF CHINA

Wu Wangshi
(Nanjing Institute of Geology and Palaeontology, Academia Sinica)

TECTONIC SETTING OF THE CARBONIFEROUS

The Carboniferous of China, formed in the middle phase of the Hercynian orogeny in Late Palaeozoic, is mainly characterized by the basical establishment of the Asian Old-land configuration with late Early Carboniferous as the enforcement phase of the Hercynian orogenetic movement. In this phase the principal geological tectonic units include: North China Continental Margin Domain, South China Continental Margin Domain and East China Continental Domain.

SEDIMENTARY STRATIGRAPHICAL UNITS AND BIOSTRATIGRAPHICAL BOUNDARY

The sediments of Carboniferous in China could be divided into four types:

The first type is called the normal neritic type which is essentially distributed in South China, and mainly consists of carbonate limestones, uaually 1 000-1 500m in thickness with more or less complete sequences. The flourished biota is characterized by the benthonic organisms.

The second type is a littoral coal-bearing type characterized by coal series of alternation of marine and continental deposits. The Lower Carboniferous coal series with less thickness is restrictly developed in South China, while the coal series of Upper Carboniferous is widely distributed in North China.

The third type--volcanic type was developed in the geosynclinal regions. It consists mainly of marine clastics with volcanic lava and tuff, and is about 5 000m in thickness. The Carboniferous sections are often possessed of local unconformity and depositional gaps, yielding the fossils which are mainly brachiopods and corals, and ammonoids as common elements. The deposits of this type were mainly developed around the Tianshan-Hinggan area.

The last type, the intermedium type, was developed in the area between stable domains and unstable domains such as the Qilianshan and Kunlun regions. The deposits of this type are complex, including carbonate, clastic and various kinds of volcanic rocks.

On the basis of fossils and sedimentary characters and biogeographical provinces of the Carboniferous System of China, there are 6 stratigraphical regions for the China's

Carboniferous System, namely: Tianshan-Hinggan region (I); North China Region (II); Qilianshan-Qinling Region (III); North Xizang-West Sichuan Region (IV); Himalaya-West Yunnan Region (V); and South China Region (VI) (Fig. 1).

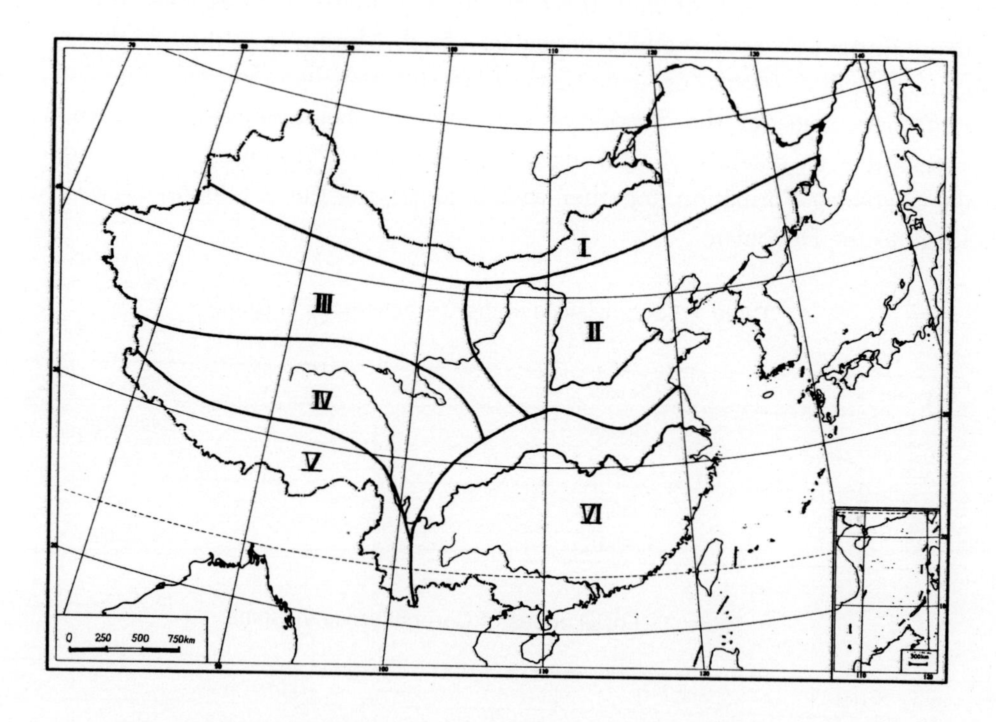

Fig. 1 Map showing the stratigraphic region of Carboniferous in China

I. Tianshan-Hinggan Region; II. North China Region; III. Qilianshan-Qinling Region; IV. North Xizang-West Sichuan Region; V. Himalaya-Yunnan Region; VI. South China Region.

The Carboniferous System of China has conventionally been divided into three series. In recent years, based on the researches of biotic and lithological characters and palaeogeographical provinces of the Carboniferous of China, especially of South China, it has been proposed to be divided into two series, and furtherly into four stages in ascending order, as: the Aikuanian, the Tatangian, the Weiningian and the Mapingian (Table 1). The typical fossil zones of each stage are recognized correspondingly as (Table 2): *Cystophrentis-Plicatifera tenuistrata* Assemblage Zone, *Pseudouralinia-Martiniella* Assemblage Zone, *Kueichouphyllum sinense-Vitiliproductus groberi* Assemblage Zone, *Pseudostaffella* Zone, *Profusulinella* Zone, *Fusulina-Fusulinella* Zone, *Montiparus* Zone, *Triticites* Zone, and *Pseudoschwagerina* Zone.

In China, the Devonian-Carboniferous boundary strata are represented by two kinds of the biotic and sedimentary types, the shallow water coral-brachiopod facies and deep water ammonoid-conodont facies. The deposits of the shallow water facies are called the Shaodong Formation, which is extensively distributed in South China, while those of deep water ammonoid-conodont facies are limited in a very narrow area. According to the definition for the boundary recommended by the International Devonian-

Carboniferous Boundary Working Group, the appearance of *Siphonodella sulcata* should be taken as the lowermost boundary for the Carboniferous. Such a boundary can be recognised in the Muhua Section of Guizhou and the Nanbiancun section in Guilin of Guangxi. But the problem of how to correlate such kinds of boundary strata with the Shaodong Formation so far still remains unsolved. Moreover, although the Shaodong Formation is rich in fossils, palaeotologists have different opinions on its geological age. Some of them consider the Shaodong Formation as belonging to the Carboniferous, others regard it as diachronous through Devonian and Carboniferous, still others hold that the Shaodong Formation, together with its overlying the *Cystophrentis*-bearing strata belongs to the Devonian.

Table 1 Subdivisions of the Carboniferous in China

System	Series	Stage
Carboniferous	Hutian (U. Carb.)	Mapingian
		Weiningian
	Fengning (L. Carb.)	Tatangian
		Aikudanian

Table 2 Fossil zones of Carboniferous in China

System	Series	Stage	Fossil zone
Carboniferous	Hutian	Mapingian	*Pseudoshchwagerina* Zone
			Triticites Zone
			Montiparus Zone
		Weiningian	*Fusulina - Fusulinella* Zone
			Profusulinella Zone
			Pseudostaffella Zone
	Fengning	Tatangian	*Yuanophyllum kansuense - Gigantoproductus edelburgensis* Assemblage Zone
			Kueichouphyllum sinense - Vitiliproductus groberi Assemblage Zone
		Aikuanian	*Pseudouralinia - Martiniella* Assemblage Zone
			Cystophrentis - Plicatifera tenuistrata Assemblage Zone

During the recent several years, some of mid-Carboniferous boundary sections have been successively discovered from such places as Baping, Nandan of Guangxi, Luodian of Guizhou, Jingyuan of Gansu. Among them, the exposure of the section in Luodian of Guizhou is excellent, which contains such fossils as the conodont *Declinognathodus,* the foraminifer *Globivalvulina moderata,* etc.

In China, the Carboniferous-Permian boundary is not consistent with the current usage abroad. Over a long period of time, the *Pseudoschwagerina* Zone has been conventionally placed in the Carboniferous in China. In recent years, after making independent studies on the evolution of fusulinids, corals and ammonoids of the Carboniferous and Permian, the Chinese palaeontologists and stratigraphers have put forth different schemes, which are as many as 4 or 5 in number, for the Carboniferous-Permian boundary in China. To sum up, there are mainly two different ideas about the attribution of the *Pseudoschwagerina* Zone. Based on the study of the evolution of fusulinids, some people considered that *Pseudoschwagerina* Zone still should be attributed to the Late Carboniferous. However, judging from the characteristics of ammonoids and corals, others have proposed that the *Pseudoschwagerina* Zone should be attributed to the Early Permian. Up to now, there had not been any scheme for the Carboniferous-Permian boundary which can be accepted by most palaeontologists and stratigraphers in China.

REGIONAL STRATIGRAPHY

1. South China Region (Fig. 2)

The Carboniferous rocks are distributed very extensively, with excellent outcrops, a great variety of fossil groups and other carbonates. The rocks mostly composed of bioclastic limetstones with a total thickness of about 1 000 to 1 500m. They belong to the platform type. In the early and middle stages of the Early Carboniferous, there are littoral or alternating marine and continental deposits intercalated with coal in some places, whereas the Carboniferous deposits of the rest stages are all carbonat sediments of shallow platform. There are various kinds of fossils mainly represented by benthonic organisms, with a high diversity both in genera and in species with a rapid evolution. This region has the earliest research history, with its biostratigraphy being studied in greatest detail, and therefore can be taken as the basis for series division and stage-establishment for the Carboniferous System of China.

Through studies over more than half a century, the zonation of different fossil groups in the Carboniferous of the South China Region have been well established. As early as at the beginning of this century, the Early Carboniferous of Guizhou was first divided in ascending order into the Kalaoho, Tangpakou, Chiussu, and Shangssu Formations. The Late Carboniferous Weining Formation and Maping Formation were established in west Guizhou and central Guangxi respectively. A Baizuo Formation has been separated from the upper part of the Shangssu Formation. However, in some local areas of South China, depositional characteristics are rather complicated, with greater changes in lithological facies. For example across the transverse section from Zhijin, Guizhou to Nandan, Guangxi through Changshun (in Guizhou), the Carboniferous deposits grade from littoral and neritic facies into deep water facies with a biota mainly composed of benthonic ones such as corals and brachiopods gradually changing to a biota mainly composed of planktonic ones such as ammonoids and conodonts. It is worth noticing that the sediment facies and biotas lead some people to infer that this region probably had been a deep trough in the ocean or a deep depression on the carbonate platform.

System	Series	Formation	Lithology	Thick. (m)	Fossils
Carboniferous	Upper Carboniferous	Maping Fm.		160	*Quasifusulina longissima* *Triticites pusillus* *Pseudocarniaphyllum orientale*
		Weining Fm.		250	*Fusulina bocki* *Pseudostaffella sphaeroidea* *Profusulinella parva* *Kionophyllum dibunum*
	Lower Carboniferous	Baizo Fm.		150	*Palaeosmilia murchisoni* *Aulina rotiformis* *Gigantoproductus edelburgensis*
		Shangssu Fm.		250 | 400	*Kueichouphyllum heishihkuanense* *Yuanophyllum kansuense* *Delepinea comoides* *Balahonia yunnanensis*
		Chiussu Fm.		300	*Kueichouphyllum sinense* *Thysanophyllum shaoyangense* *Vitiliproductus groberi* *Megachonetes zimmermanni*
		Tangpakou Fm.		160	*Pseudouralinia gigantea* *Eochoristites neipentaiensis* *Martiniella chinglungensis*
		Kolaoho Fm.		150	*Cystophrentis kolaohoensis* *Schuchertella gueizhouensis* *Composita globosa*

Fig. 2 Carboniferous of South China (Guizhou Province)

2. North China Region (Fig. 3)

This region is devoid of Early Carboniferous deposits. The Late Carboniferous sediments, for example, the Carboniferous deposits of Taiyuan, Shanxi belong to the littoral alternating marine and continental facies, with a series of exploitable coal-seams. The early Late Carboniferous strata are called the Penchi Formation, with a small thickness of some twenty meters only, while the late ones are named the Taiyuan Formation, about more than 100m in thickness. Besides animal fossils these rocks also contain rich fossil plants, which include not only elements of the Cathaysian flora, such as *Cathaysiodendron incertum, Tingia hamaguchii* and *Lepidodendron oculis-felis*, but also some cosmopolitan elements as well, such as *Lepidodendron posthumi* and *Neuropteris ovata.*

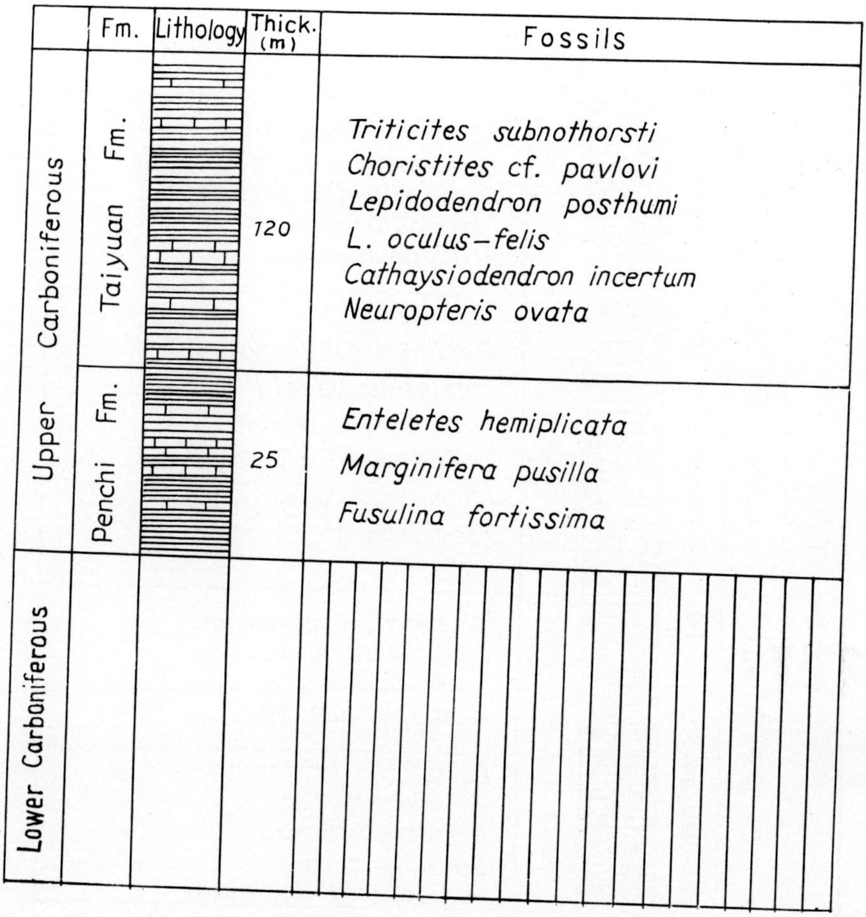

	Fm.	Lithology	Thick. (m)	Fossils
Upper Carboniferous	Taiyuan Fm.		120	*Triticites subnothorsti* *Choristites* cf. *pavlovi* *Lepidodendron posthumi* *L. oculus-felis* *Cathaysiodendron incertum* *Neuropteris ovata*
	Penchi Fm.		25	*Enteletes hemiplicata* *Marginifera pusilla* *Fusulina fortissima*
Lower Carboniferous				

Fig. 3 Carboniferous of North China (Shanxi Province)

3. Tianshan-Hinggan Region (Fig. 4)

The Carboniferous rocks in this region have a much greater thickness which is generally more than 5 000m, with intercalations of various kinds of volcanic rocks and marine clastic rocks together with carbonates, and local unconformity and depositional gaps often occurred. In addition to the benthonic organisms deposits, there are ammonoids, and conodonts of the planktonic facies. All these indicate that the deposits of the region belong to the geosyncline type. In recent years, in the Carboniferous section along the northwestern margin of the Junggar Basin, the Carboniferous System exceeds 8 000m in total thickness, which is divided in ascending order into the Lower Series, including the Hebukehe Formation, the Xibeikulasi Formation and the Baogutu Formation, and the Upper Series, including the Halaalate Formation and the Aladeyikesai Formation. The Lower Series is characterized by the *Syringothyris*, with abundant radiolarian *Cenasphaera*, the ammonoid *Gattendorfia* together with different species of the conodont *Siphonodella* at the basal part. In the Upper Series, a considerable number of fusulinids and brachiopods have been discovered which are similar to those of South China.

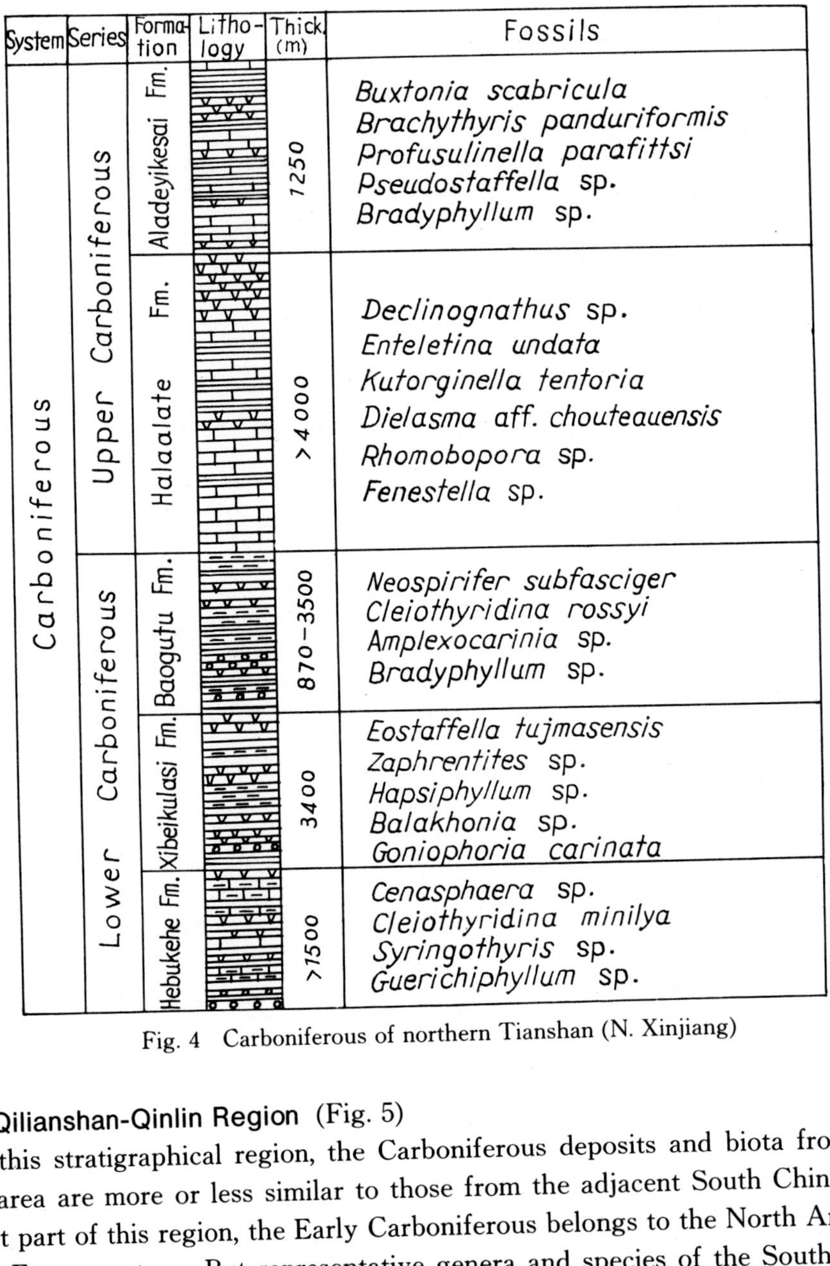

System	Series	Formation	Lithology	Thick. (m)	Fossils
Carboniferous	Upper Carboniferous	Aladeyikesai Fm.		1250	*Buxtonia scabricula* *Brachythyris panduriformis* *Profusulinella parafittsi* *Pseudostaffella* sp. *Bradyphyllum* sp.
		Halaalate Fm.		>4000	*Declinognathus* sp. *Enteletina undata* *Kutorginella tentoria* *Dielasma* aff. *chouteauensis* *Rhomobopora* sp. *Fenestella* sp.
	Lower Carboniferous	Baogutu Fm.		870–3500	*Neospirifer subfasciger* *Cleiothyridina rossyi* *Amplexocarinia* sp. *Bradyphyllum* sp.
		Xibeikulasi Fm.		3400	*Eostaffella tujmasensis* *Zaphrentites* sp. *Hapsiphyllum* sp. *Balakhonia* sp. *Goniophoria carinata*
		Hebukehe Fm.		>1500	*Cenasphaera* sp. *Cleiothyridina minilya* *Syringothyris* sp. *Guerichiphyllum* sp.

Fig. 4 Carboniferous of northern Tianshan (N. Xinjiang)

4. Qilianshan-Qinlin Region (Fig. 5)

In this stratigraphical region, the Carboniferous deposits and biota from the south border area are more or less similar to those from the adjacent South China Region. In the west part of this region, the Early Carboniferous belongs to the North American type and the European type. But representative genera and species of the South China type, such as ***Cystophrentis, Pseudouralinia, Eochoristites*** and ***Martiniella*** have not been discovered therein so far. The Carboniferous sections of this region can be represented by those along the south slope of Qilianshan. In this area, the Carboniferous strata are about 2 000m in thickness, the lower part of the Lower Series consists of conglomerate rocks. The lower part of the Upper Series is similar to the Penchi Formation in North China, consisting of alternating marine and continental strata with intercalations of coal. Other beds and members are all composed of carbonate deposits. The stratigraphical sequence had been recognized in ascending order as the Chengchiangou and Huaitoutala

Formations of the Lower Series and the Keluke and Maping Formations of the Upper Series. The Chengchiangou Formation yields *Vesiculophyllum* of the North American type and *Kassinella* from Kazakhstan, the interesting materials discovered in recent years.

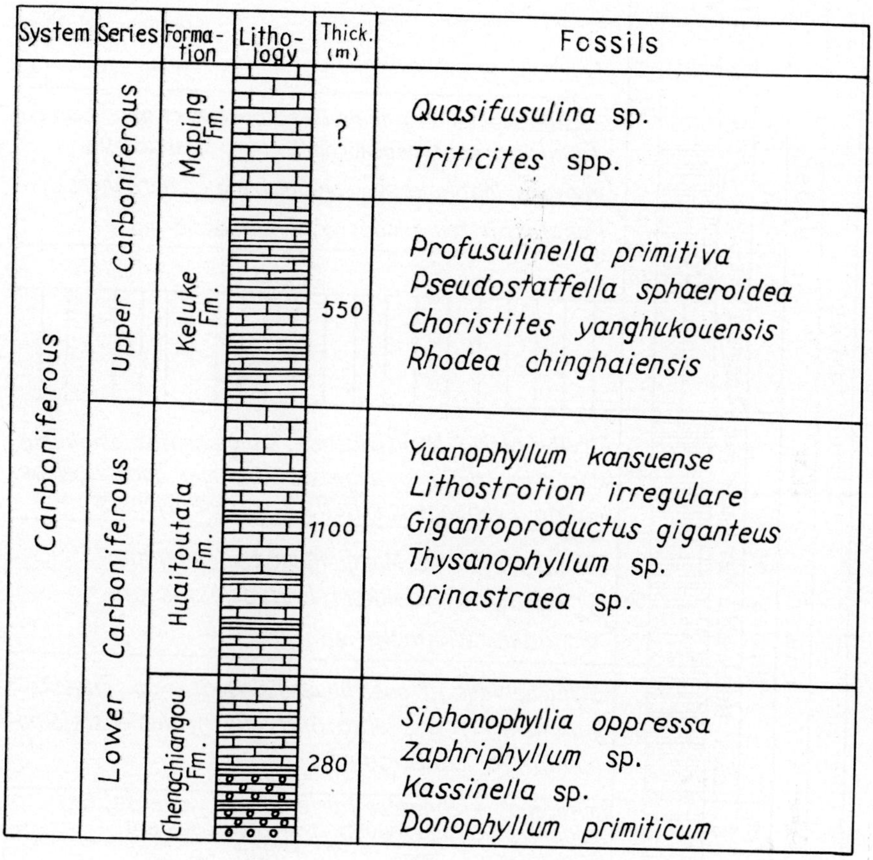

System	Series	Formation	Lithology	Thick. (m)	Fossils
Carboniferous	Upper Carboniferous	Maping Fm.		?	*Quasifusulina* sp. *Triticites* spp.
		Keluke Fm.		550	*Profusulinella primitiva* *Pseudostaffella sphaeroidea* *Choristites yanghukouensis* *Rhodea chinghaiensis*
	Lower Carboniferous	Huaitoutala Fm.		1100	*Yuanophyllum kansuense* *Lithostrotion irregulare* *Gigantoproductus giganteus* *Thysanophyllum* sp. *Orinastraea* sp.
		Chengchiangou Fm.		280	*Siphonophyllia oppressa* *Zaphriphyllum* sp. *Kassinella* sp. *Donophyllum primiticum*

Fig. 5 Carboniferous of Qilianshan (Qinghai Province)

5. Himalaya-West Yunnan Region (Fig. 6)

In this region, the Carboniferous strata are mainly composed of clastic rock intercalated with limestone about 2 500m in thickness. They can be divided into the Yali and Naxing Formations of the Lower Series and the Jilong Formation of the Upper Series. Here it must be pointed out that the lower part of the Yali Formation possibly belongs to the Devonian. The upper part of the Yali Formation yields *Gattendorfia* and *Imitoceras,* as well as *Pseudosyrinx* and *Syringothyris* of the North American type. Up to the present time, no reliable early Late Carboniferous sediment has been discovered. The Jilong Formation, representing deposits of the Late Carboniferous, consists of glaciomarine diamictites, yielding a fauna characterized by *Stepanoviella,* which are similar to the *Stepanoviella* fauna of Kashmir and the Umaria Bed of India. Based on the lithological and biological features of the Jilong Formation, the Carboniferous of this region is considered to be of Gondwana type.

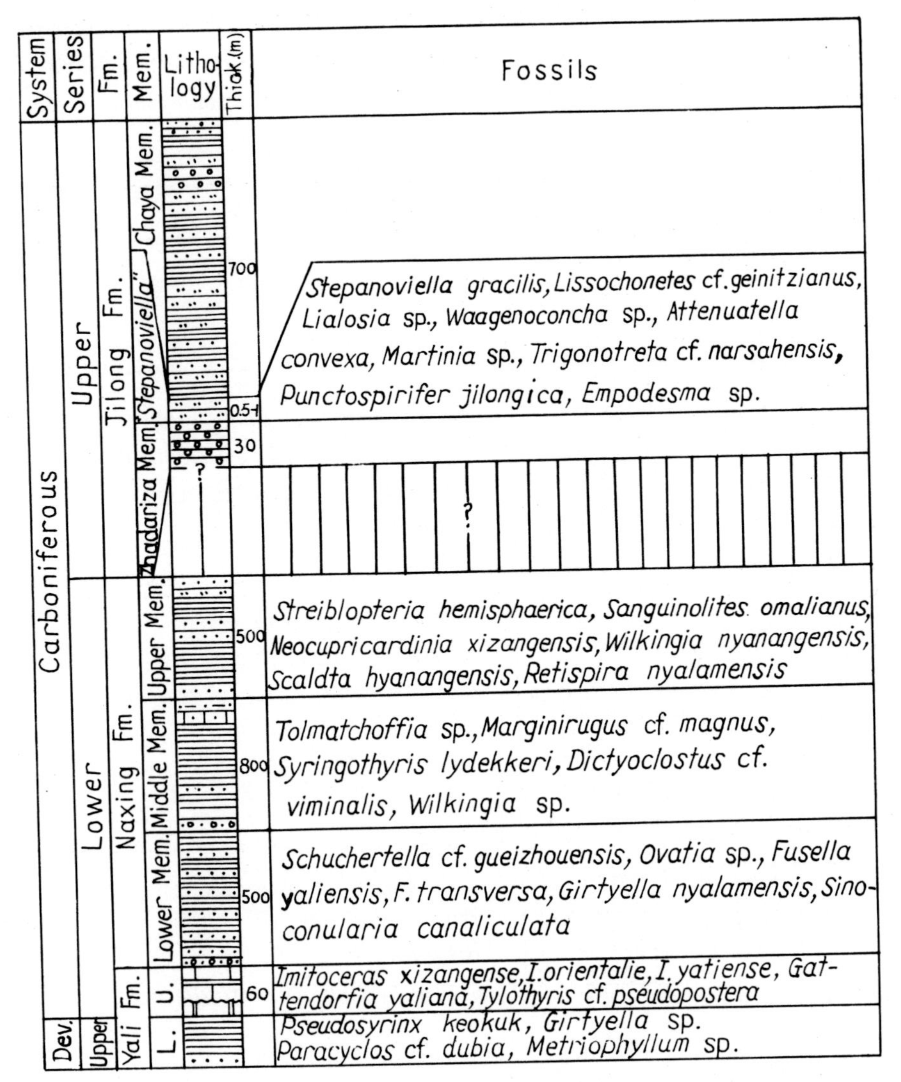

Fig. 6 Carboniferous of Himalaya area

As a matter of fact, in the area of west Yunnan, even part of west Sichuan, the Carboniferous sediments and biotas are also relatively complicated in character, containing not only deposits and biotas of the Gondwana type, but also some fossils of the South China type as well. This phenomena may greatly interest tectonic geologists.

6. North Xizang -West Sichuan Region (Fig. 7)

This region represents the north facies of the Carboniferous sediments in Xizang. Taking the Qamdo section for example, the Carboniferous sequence can be recognized in ascending order as: Lower Series, including the Wuqingna and Machala Formations, and the Upper Series, including the Aoqu and Licha Formations. Lithologically, this section is entirely different from those of the South Gondwana facies. It is rather similar to the Carboniferous of the South China Region in lithological characters, i.e. mainly composed of bioclastic limestones, and intercalated with cherty limestones, more than 2 500m in thickness. Besides the similarity in lithological characters, the fossil contents are also extremely similar. For example, the characteristic elements of the South China Region also occur in this section, such as *Cystophrentis, Eochoristites, Kueichouphyllum, Gondolina,* etc.

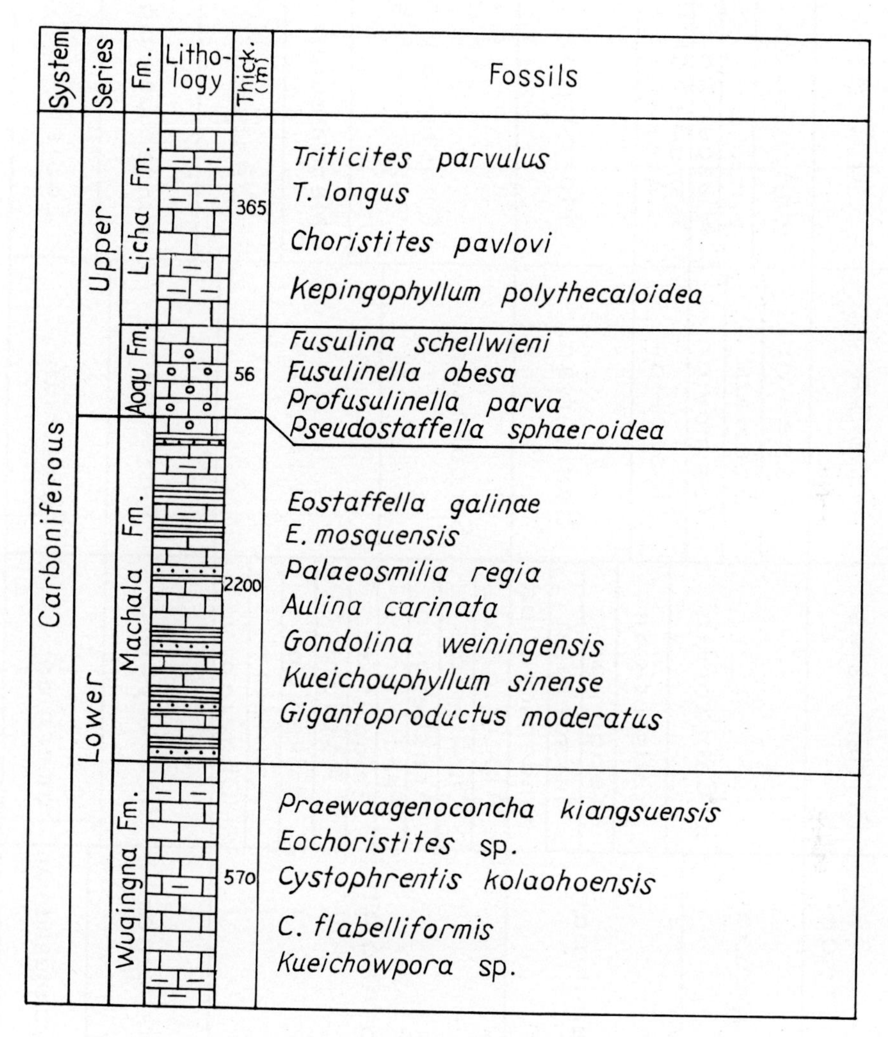

Fig. 7 Carboniferous of N. Xizang and W. Sichuan (Qamdo area)

CORRELATION OF THE CARBONIFEROUS

Correlation of the Carboniferous System in China with contemporaneous strata in other countries is shown in Table 3.

1. The establishment of the four stages in China depends on material from South China. The Aikuanian Stage of the Early Carboniferous in South China is characterized by the occurrence of endemic elements. Therefore, there exist some difficulties in direct correlations with the contemporaneous strata in Britain, USSR and USA, but there are still some similarities.

Table 3 Correlation of Carboniferous subdivision in China and other countries

China: Series	China: Stage	British Isles	USSR	USA
Hutianian (U. Carb.)	Mapingian	L. Cantabrian	Asselian	Wolfcampian
			Gzelian	Virgilian
			Kasimovian	Missourian
	Weiningian	Radstockian	Moscovian	Desmoinesian Atokan
		Yeadonian	Bashkirian	Morrowan
		Marsdenian		
		Kinderscoutian		
Fengningian (L. Carb.)	Tatangian	Alportian	Serpukhovian	Chesterian
		Chokierian		
		Arnsbergian		
		Pendleian		
		Brigantian	Visean	
		Asbian		Meramecian
		Holkerian		
		Arundian		Osagean
		Chadian		
	Aikuanian	Courceyan	Tournaisian	Kinderhookian
Xikuangshanian (U. Dev.)	Xikuangshanian	Famennian	Famennian	Louisiana

2. The Courceyan Stage of Britain yields the corals *Caninia pustula* and *Siphonophyllia,* which also occur in the Aikuanian Stage. Both can be correlated.

3. Based on the appearances of the corals *Lithostrotion, Dibunophyllum, Diphyphyllum* and *Orionastrenea* and the brachiopods *Delepinea* and *Davisiella* both in the Tatangian Stage of China and in the Visean Stage of Britain, these two stages can also be correlated with each other.

4. From the Carboniferous sections in Shuicheng of Guizhou, there have discovered the ammonoids *Homoceras, Homoceratoids, Retites, Gastrioceras* and *Branneroceras;* this proves that the corresponding members the Namurian of West Europe and the Serpukhovian of USSR, are developed in China.

5. In West Europe, the Late Carboniferous strata are wholly continental coal-bearing deposits, which are similar to those in North China. The South China Sea appears to be more closely related with the European Sea of USSR; both can be easily correlated with each other. The Mapingian in China covers a very large scheme, and it can be correlated approximately with the Kasimovian, Gzhelian and Asselian of the Russian Platform.

SUMMARY

1. In South China, further investigations in greater detail will be made into the chronological strata of the Carboniferous System.

2. Biostratigraphical work will be further carried on with regard to the Carboniferous System around the marginal and geosynclinal areas of China.

3. In view of the excellent outcrops in China, contineous strata and rich fossils of the Carboniferous, there are certain favourable conditions in which for us to establish standard sections of boundary in order to make international correlation with those contemporaneous strtata throughout the world.

4. To raise the level of researches on the Carboniferous of China, it is necessary to mark further studies on the interrelationship between the biological facies and lithological facies of the Carboniferous in China, and to carry on multi-disciplinary and comprehensive studies of biostratigraphy, palaeogeomagnetic and organic geochemistry.

REFERENCES

1. Hou Hongfei, Wang Zengji et al, The Carboniferous System of China, Geological Publishing House (1982).
2. Jing Yukan and Sun Dongli, In: Palaeontology of Xizang, III, Science Press (1981), 127.
3. Wang Zengji, Geol Rev, 27 (6) (1981), 533.
4. Wang Hongzhen (ed), In: Atlas of the Palaeogeography of China, Cartographic Publishing House (1985).
5. Wu Wangshi et al, In: Palaeontology of Xizang, V, Science Press(1982), 107.
6. —— and Kong Lei, Palaeontologia Cathayana, 1 (1987), 367.
7. —— et al, Acta Palaeont Sinica, 23(4) (1984), 411.
8. Yang KC, Sci Rec, 4(4) (1960), 185.

9. Yang Jingzhi, Wu Wangshi, Zhang Linxin et al, In: Classification and Correlation of the Carboniferous in China, Science Press (1982).
10. Zhang Linxin, Acta Palaeont Sinica, 24(2) (1985), 205.
11. Zhao Xiuhu et al, Bull Nanjing Inst Geol Palaeont, Acad. Sinica, 5 (1982), 1.

XIe Congrès International de Stratigraphie et de Géologie du Carbonifère Beijing 1987,
Compte Rendu 1 (1991): 97-116

THE TECTONO-PALAEOGEOGRAPHY AND BIOGEOGRAPHY OF CHINA AND ADJACENT REGIONS IN THE CARBONIFEROUS PERIOD

Wang Hongzhen
(China University of Geosciences, Beijing)
Zheng Luiren
(Ministry of Geology and Mineral Resources, China)
Wang Xunlian
(China University of Geosciences, Beijing)

This paper contains three parts. The first part is a brief review of the geotectonic background, the tectono-palaeogeography of China and adjacent regions in the Late Palaeozoic Hercynian Stage. The second part deals with the biogeography of China and of the world, with special emphasis on rugose corals. The third part includes a brief account of the palaeogeography and some related sedimentary mineral deposits in China.

TECTONIC FRAME OF CHINA AND ADJACENT REGIONS IN THE HERCYNIAN STAGE

According to the mobilist view on global tectonics, and the idea of development by stages in regard to crustal evolution,[1,2] the spatial differentiation of crustal sectors permits a classification of three ranks of tectonic units: The first rank is the "tectonic domain" which may be defined as a group of continental cratons and their continental margins including microcontinents or massifs and fold belts. For example, the North Asian domain contains Siberia, Kazakhstania and the Mongolia-Hinggan Palaeozoic fold belts, with included small massifs. The second rank "subdomain" includes two types. The stable or mature type of subdomain consists of a main continental craton and its attached marginal tracts, and is exemplified by the Sino-Korean subdomain, which includes not only the main craton, but also the Qaidam and Longxi massifs in the Qilian Caledonian fold belt. The second or mobile type is generally composed of a group of separate yet mutually related massifs connected by fold belts, as exemplified by the Mongolia-Hinggan subdomain which contains Kuznetzk, Sayan Ubsu, Bayan Honger massifs and Palaeozoic fold belts. The third rank comprises the traditional tectonic units platforms and fold belts and need not be further discussed.

The main thought of this classification is to emphasize the continental margin tracts, which are affiliated or attached to the major continental cratons, and the integrated entity may be regarded as the geotectonic units of higher rank. Based on these concepts, the crustal consumption zones or geosutures making up the boundary between the geotectonic units, as generally recognized in the mobilist view, may be classified into two kinds. Those separating the tectonic domain or subdomain are the more important convergent zones, and those delineating the fold belts within one subdomain, or usually within one continental margin tract, are the accretional zones. The possible development of these two kinds of geosutures was explained in a diagram showing the convergence of two opposite continental margins,[3] which includes the successive subduction and accretion of the continental margins to their mainlands respectively, finally resulting in the convergence and collision of the opposite continents. In regard to development of crustal evolution, we have attempted to show that this procedure of development (which we have called the complication of continental margins) may have been of characteristic in an important stage, the third megastage of the formation of Pangaea, covering the time span of from 800 to 200Ma, i.e. from the Sinian to the Triassic.[1,2,3] It may be pointed out that, these geotectonic considerations are helpful, and sometimes are essential in understanding and formulating the global tectonic frame and biogeographical outlines. Part of the domain and subdomains are indicated on the Hercynian tectono-palaeogeographic map (Fig. 1). The main convergent zones, now separating the North Asian, the Central Asian, the South Asian and Gondwanaland, including the subsequent, post-Hercynian sutures within the South Asian and the Gondwanan domains, are also shown. As will be shown in the following, they have an important bearing on biogeographic as well as in geotectonic analysis.

In the Hercynian map (Fig. 1), we have specially emphasized three aspects. The first is the nature of the continental margins, especially the distribution of island arcs and ophiolite suites. There are three pronounced island arc belts. The first belt is in the northwest and includes three crustal consumption zones. From north to south, they are: 1) the early Hercynian convergent zone along the Irtysh, between Salair Sayan and Kazakhstan, continuing eastwards to the south of Bayan Honger and of Tuotuoshan (Fig. 1); 2) The early Hercynian accretional zone, northerly subducted along the northern Tianshan between the Junggar and the Yining massifs; and 3) the late Hercynian convergent zone along northern Tianshan and Beishan. This last zone is the most important boundary between the North Asian domain and the Central Asian domain, which was however not completely closed until in Late Permian. The second important island belt starts from West Kunlun, and includes the West Kunlun and the East Kunlun. A northerly subducted accretional zone is situated to the south of Tarim and of Qaidam. Further to its south is the main convergent zone separating the Sino-Korean, including the Qaidam, in the north, and the Yangtze-Qiangtang in the south, which was closed in the Indosinian stage, and forms the principal boundary between the Central Asian domain and the South Asian domain. The third belt lies in the Sanjiang region, and is especially clear in west Yunnan along the western border of the Lincang massif, with an eastward subduction of Carboniferous age. It may have continued to the north, in a zone between the Qamdo massif in the east and the eastern part of the Qiangtang massif in the west. This zone probably marks the Palaeozoic boundary between the South Asian domain and the Gondwana domain, but later movements have caused repeated displacements of the related terrains and obliterated the original relations between them. In these belts the ultrabasic rocks representing possible residual oceanic crust and granite zones formed of later collision are also indicated. The hatchered zones on the map denote the probable sites of Late Palaeozoic continental slopes facing open oceanic basins at the time.

The second aspect to be emphasized is the Carboniferous tensile structures developed on the platforms and on the Caledonides, which are represented by the basic mantle-type volcanism and the microaulacogens.[3] These are known in the Yining massif where Late Carboniferous albitophyre and trachyandesites of aulacogen type abound, and near the northern border of Tarim, where Late Carboniferous deep water deposits occur in juxtaposition with neritic deposits. The pronounced dark sediments belt from Ziyun to Shuicheng in Guizhou, cutting in the Upper Yangtze craton and ranging from Devonian to Permian in age, is also well known.[3] Carboniferous micro-aulacogens are also reported from northern Guangxi, western Fujian[4] and in the Loping area in northeastern Jiangxi, all situated on the Caledonides.

Finally, the third aspect of emphasis refers to the extensive inland type of near-shore and terrestrial basins within the scope of the stable cratons. They were originated either as a result of lateral compression, as in North China and Mongolia, or of large scale rifting and faulting, as in India, in the late Hercynian to post-Hercynian time. The formation of large scale intracratonic terrestrial basin of Late Carboniferous to Permian may have been a peculiar and unprecedented phenomenon in the sedimentary history of the earth.

EARLY CARBONIFEROUS BIOGEOGRAPHY OF CHINA AND OF THE WORLD

In this paper the Early Carboniferous biogeography of China and of the world are mainly based on analysis of rugose corals and also of brachiopods. The essential point in biogeographical analysis is that the taxonomic units used in calculation should be genuinely on the same taxonomic level. In recent years we have made SEM studies on the micro-skeletal structures of rugose corals, altogether more than 80 genera belonging to almost all the important families having been investigated. Based on this and other considerations, we have made an overall revision of the subclass Rugosa, having recognized altogether 795 genera classified into 5 orders, 13 suborders and 79 families.* The Lower Carboniferous contains 183 rugose genera belonging to 39 families, which will, we consider, provide a rather sound basis for biogeographic analysis and comparison.

In regard to biogeographical classification, we have used a three ranks category, realms, provinces and subprovinces. While it is not necessary to enter into the definition of these terms, it may be adequate to point out that the main controlling factor in biogeographic differentiation is, according to the ideas of variance of biogeography and

* to be published in the monograph entitled *Classi fication and Evolution of the Palaeozoic Corals of China.*

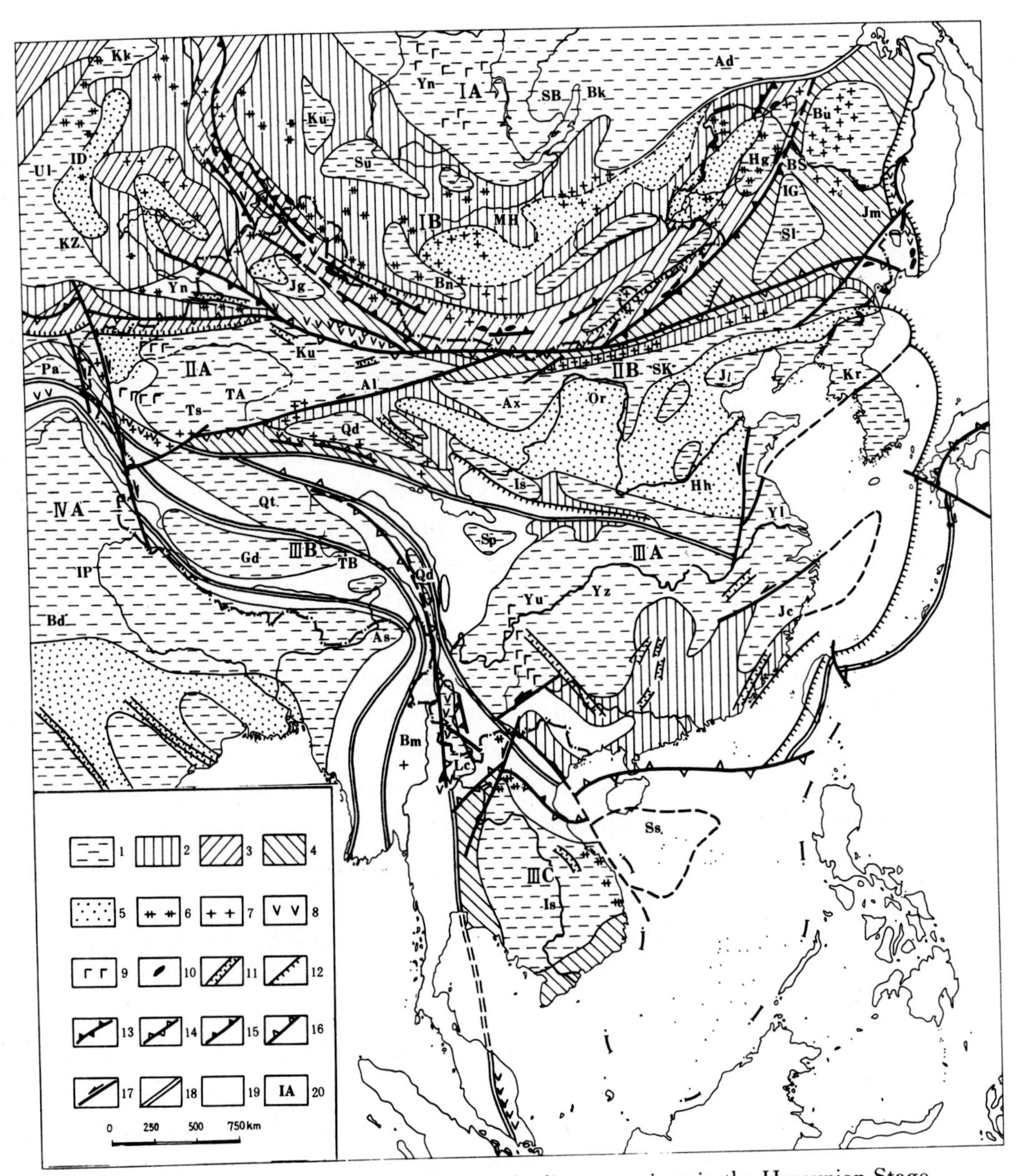

Fig. 1 Palaeotectonic map of China and adjacent regions in the Hercynian Stage

1. Pre-Sinian paltforms and massifs; 2. Caledonides; 3. Early Hercynides; 4. Late Hercynides; 5. Mainly terrestrial Permian-Triassic basins on cratons; 6. Early Hercynian granites; 7. Late Hercynian granites; 8. Hercynian island-arc volcanics; 9. Permian-Triassic traps; 10. Ultrabasic rocks; 11. Late Palaeozoic aulacogens and micro-aulacogens; 12. Late Palaeozoic continental slopes; 13. Early Hercynian convergent crustal consumption zones; 14. Late Hercynian convergent crustal consumption zones; 15. Early Hercynian accretional crustal consumption zones; 16. Late Hercynian accretional crustal consumption zones; 17. Post-Hercynian transcurrent fault zone; 18. Post-Hercynian crustal consumption and collision zones; 19. Post-Hercynian fold belts and terrains; 20. Geotectonic units.

Geotectonic units I NORTH ASIAN (ANGARACHSTAN) DOMAIN IA SB Siberian subdomain; Ad Aldan nucleus; Yn Yenesei belt; Bk Baikal belt. IB MH Mongolia-Hinggan subdomain; Ku Kuznetzk massif; Su Sayan-Ubsu massif; Bh Bayan-Honger massif; Tt Tuotuoshan massif; Hg North Hinggan massif. IC BS Bureya-Songliao subdomain; Bu Bureya massif; Jm Jamus massif; Sl Songliao massif. ID KZ Kazakhstan subdomain; Kk Kokchitav massif; Ul Ulutau massif; Yn Yining massif; Jg Junggar massif. II CENTRAL ASIAN (SINARIMIAN) DOMAIN IIA TA Tarim subdomain; Ts South Tarim nucleus; Qu Quruktagh belt; Al Altay belt. IIB SK Sino-Korean subdomain; Or Ordos nucleus; Jl Jiliao nucleus; Hh Hehuai nucleus; Kr Korean massif; Ax Alxa massif; Qd Qaidam massif Lx Longxi massif. III SOUTH ASIAN DOMAIN IIIA YZ Yangtze subdomain; Yu Upper Yangtze nucleus; Kd Kangdian massif; Yl Lower Yangtze massif; Jc Jian'ou-Chencai massif; Sp Songpan massif; Lc Lincang massif; Qd Qamdo massif. IIIB TB Tibet (Xizang)-Burma subdomain; Qt Qiangtang massif; Gd Gangdise massif; Hm Himalaya massif; Bm Burma massif; Ma Malaya massif; Pa Pamir massif. IIIC IS Indosinia-South Sea subdomain; Is Indosinia massif; Ss South Sea massif. IV EAST GONDWANA DOMAIN IVA IP Indian-Pakistan subdomain; Bd Bandelkhand massif; As Assam massif.

of mobilism, the isolation caused by changes in the world frame of continents and oceans. Furthermore, probably no definite figures could be fixed for provincial similarity coefficients. On the contrary, the values may change with respect to different faunal groups and different earth relief conditions in various periods. The values used in this paper for provincial distinction in the Early Carboniferous may be rather high, as faunal differentiation and isolation were not pronounced, especially in the Visean.

In the sketch map showing the Early Carboniferous biogeography of China (Fig. 2), eight areas of rugose coral faunas in China, selected respectively for the Tournaisian and the Visean, were used in a simple Otzuka similarity coefficient calculation. Representative or leading genera are used to indicate the provinces and subprovinces. The brachiopod assemblages are after Yang Shipu. Within one subprovince, for example in the Yangtze subprovince, further differentiation of endemic centers are possible and are also indicated in the map (Fig. 2). The endemic centers in general do not form a formal rank of biogeographic unit, as they are usually disseminated and discontinuous in distribution. The recognition of the Himalaya region as a part of the South Tibet (Xizang) subprovince under the South Tethyan province is preliminary, as the brachiopod faunas, the peculiar ***Dowhatania-Adminiculoria*** assemblage, shows a kinship with the Boreal Siberian forms and may represent a bipolar relationship with the latter. Thus there is the possibility that they may actually belong to the Gondwanan realm. We will return to this point in a later paragraph. It is evident that the main biogeographic boundaries coincide with the important convergent zones, as it was along these zones that vast areas of oceanic regions once present have been consummed.

In order to get a world perspective, we shall now give a brief account of the global tectonic frame and biogeography in the Early Carboniferous. In recent years, A. M. Ziegler and his group in Chicago have published a series of very fine world reconstruction maps (1979), and the Visean maps published in 1982 include some details in oceanography and phytogeography.[7] As no palaeomagnetic data of China and adjacent regions were available then, the eastern Tethys in their maps is still shown as an entire ocean. Recently we have made use of the palaeomagnetic data so far obtained in China (Table 1), and recalculations of some of the old data in other parts of the world have been made by Zhu Hong, with the purpose to revise the world reconstruction of continents, especially in the Palaeozoic. Lin has recently given a world reconstruction map of the Late Carboniferous.[8] Table 1 contains the data used for reconstruction in the Early Carboniferous map and is supplied by Zhu Hong. With an autograph program designed by Zhang Linghua and Li Xiang, and by use of an IBM PC-AT microcomputer, we have been able to produce several world reconstruction maps.

Fig. 3 is the Early Carboniferous tectono-palaeogeographic map of the world, in which a global classification of tectonic domains and subdomains has been attempted. As shown in the map, the Tethys ocean may be subdivided into a western Tethys and an eastern Tethys. The eastern Tethys includes a northern or the main part, and a southern part, which were separated by the archipelagoes of Burma-Tibet and of Iberia-Iran, both being considered as the constituent parts of the South Asian tectonic domain.

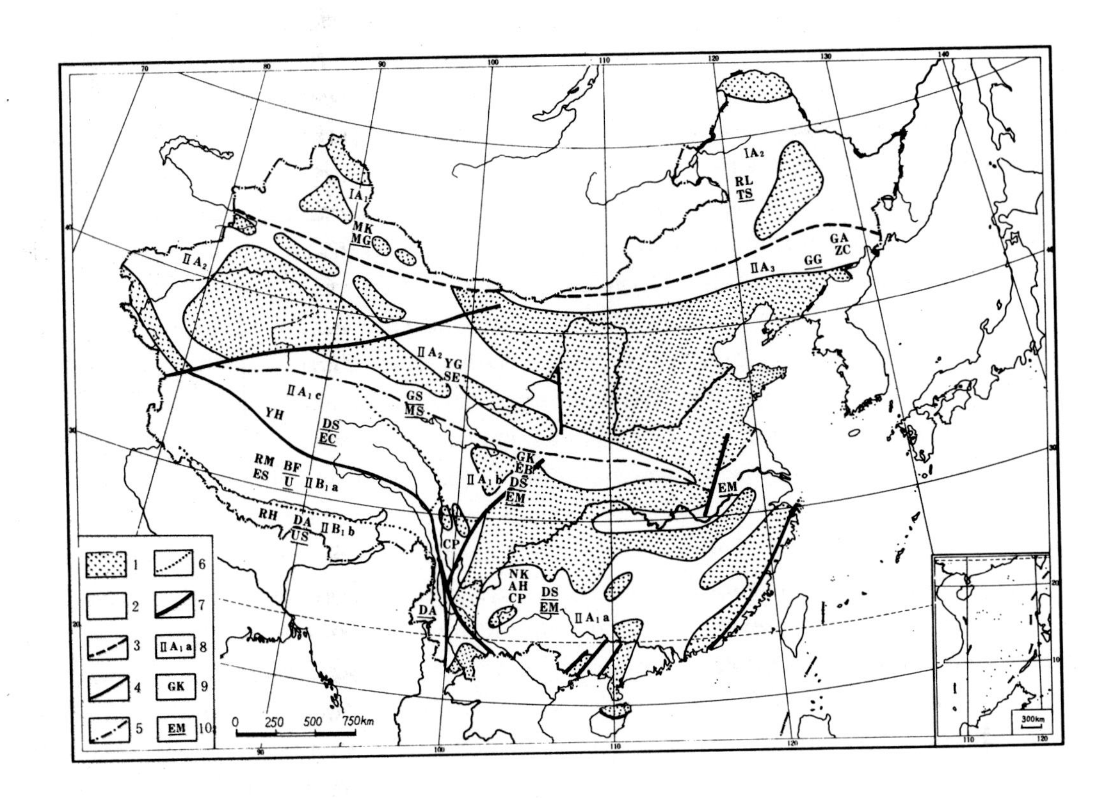

Fig. 2 Early Carboniferous biogeography of China

(based on rugose corals and brachiopods)

1. Land; 2. Sea; 3. Boundary between biogeographic realms; 4. Boundary between biogeographic provinces; 5. Boundary between biogeographic subprovinces; 6. Boundary between endemic centers; 7. Main subsequent faults; 8. Biogeographic units; 9. Rugose coral assemblages; 10. Brachiopod assemblages.

I BOREAL REALM IA Mongolia-Hinggan province IA1 Junggar-Altay subprovince Co: MK *Meniscophyllum-Kinkaidia*; Br: MG *Marginatia-Gigantoproductus*. IA2 Hinggan subprovince Co: RL *Rotiphyllum-Lithostrotionella*; Br: TS *Tolmatchoffia-Syringothyris*. II TETHYAN REALM IIA North Tethyan province. IIA1 Yangtzean subprovince IIA1a Hunan-Guizhou Co: NK *Neoclisiophyllum-Koninckophyllum*; AH *Arachnolasma-Heterocaninia*; CP *Cystophrentis-Pseudouralinia*; Br: DS *Datangia-Striatifera*, EM *Eochoristites-Martiniella*. IIA1b Longmen-Qinling Co: GK *Gangamophyllum-Kueichouphyllum*,EB *Enniskillenia-Beichuanophyllum*; Br: DS *Datangia-Striatifera*, EM *Eochoristites-Martiniella*. IIA1c Qiangtang-Tangla Co: YH *Yuanophyllum-Heterocaninia*; Br: DS *Datangia-Striatifera*, EC *Eochoristites-Camarotoechia*. IIA2 Tarim-Qinling subprovince Co: YG *Yuanophyllum-Gangamophyllum*, SE *Siphonophyllia-Enygmophyllum*; Br: GS *Gigantoproductus-Striatifera*, MS *Marginatia-Syringothyris*. IIA3 Jilin-SW Japan subprovince Co: GA *Gangamophyllum-Amygdalophyllum*, ZC *Zaphrentites-Cyathaxonia*; Br: GG *Gigantoproductus-Grandispirifer*. IIB South Tethyan province IIB1 South Tibet (Xizang) subprovince IIB1a Gangdise Co: RM *Rhopalolasma-Mirusophyllum*, ES *Ekvasophylloides-Sycnoelasma*; Br: BF *Balakhonia-Fusella*, U *Unispirifer*. IIB1b Himalaya Co: RH *Rhopalolasma-Hapsiphyllum*; Br: DA *Dowhatania-Adminiculoria*, US *Unispirifer-Syringothyris*.

The North Asian domain is also complex in structure. As no palaeomagnetic data are available, the position of the Bureya-Jamusi and northeastern Japan subdomain is hypothetical. It may be noted that the Ordovician marine faunas of this region show a kinship with those of Siberia and even with western North America, and the Tournaisian corals and brachiopods are also of Siberian type.[9] But the Visean faunas are to a certain extent similar to those of southwestern Japan and South China. Again, whether the Palaeo-Arctic was so extensive and whether it was oceanic in nature is also problematic. The relative positions of Kazakhstan, Junggar and Tarim are complicated and interesting. There seems to be little doubt that Tarim and North China (Sino-Korea) were separated from each other. The Qilian region was already accreted to the mainland of North China, but oceanic sea bottom and island arcs still existed between the Junggar, Yining and Tarim massifs. Island arcs and subduction zones are temporarily indicated on the map, as are also the possible convergent crustal consumption zones. The presence of a northerly subduction zone along the southern border of Tarim and Qaidam, which formed the northern margin of the main eastern Tethys, seems to be beyond doubt.

Table 1 Palaeomagnetic data of China used in world reconstruction of the Early Carboniferous continents

Tectonic unit	Locality	Age	Pole position		Palaeolatitude	Reference
Yangtze platform	Xichuan, Henan	C_1	61.7	254.0	10.3	Li Yanping et al. (1985)
	Longmenshan, Sichuan	C_1	51.9	299.5	5.58	Wang Zhongmin (1985)
South China Caledonian belt	N. Guangdong	C_1	80.6	357.5	20.2	Xing Yuqiu et al. (1985)
	same	C_1	72.5	329.2	10.5	same
	same	C_1	61.6	211.5	17.5	same
	same	C_{2+3}	59.6	211.7	17.3	same
	mean		72.5	261.8	14.1	
North China platform	Taiyuan, Shanxi	C_2	62.0	300.2	9.1	Lin Wanzhi et al. (1984)
	Lushan, Henan	C_2	70.4	295.8	14.3	Li Yanping et al. (1985)
	same	C_{2+3}	59.5	8.4	22.0	same
	mean		68.0	324.4	17.3	
Tarim platform	Kalpin	C_3	52.2	179.5	25.8	Bai Yuqhong et al. (1984)
Qaidam massif	Da Qaidam	C_3	35.2	310.8	10.7S	Wang Zhongmin (1985)

In regard to world rugose coral biogeography, we have listed in the Tables 2 and 3 Tournaisian and Visean genera respectively, with their distribution in the main regions in

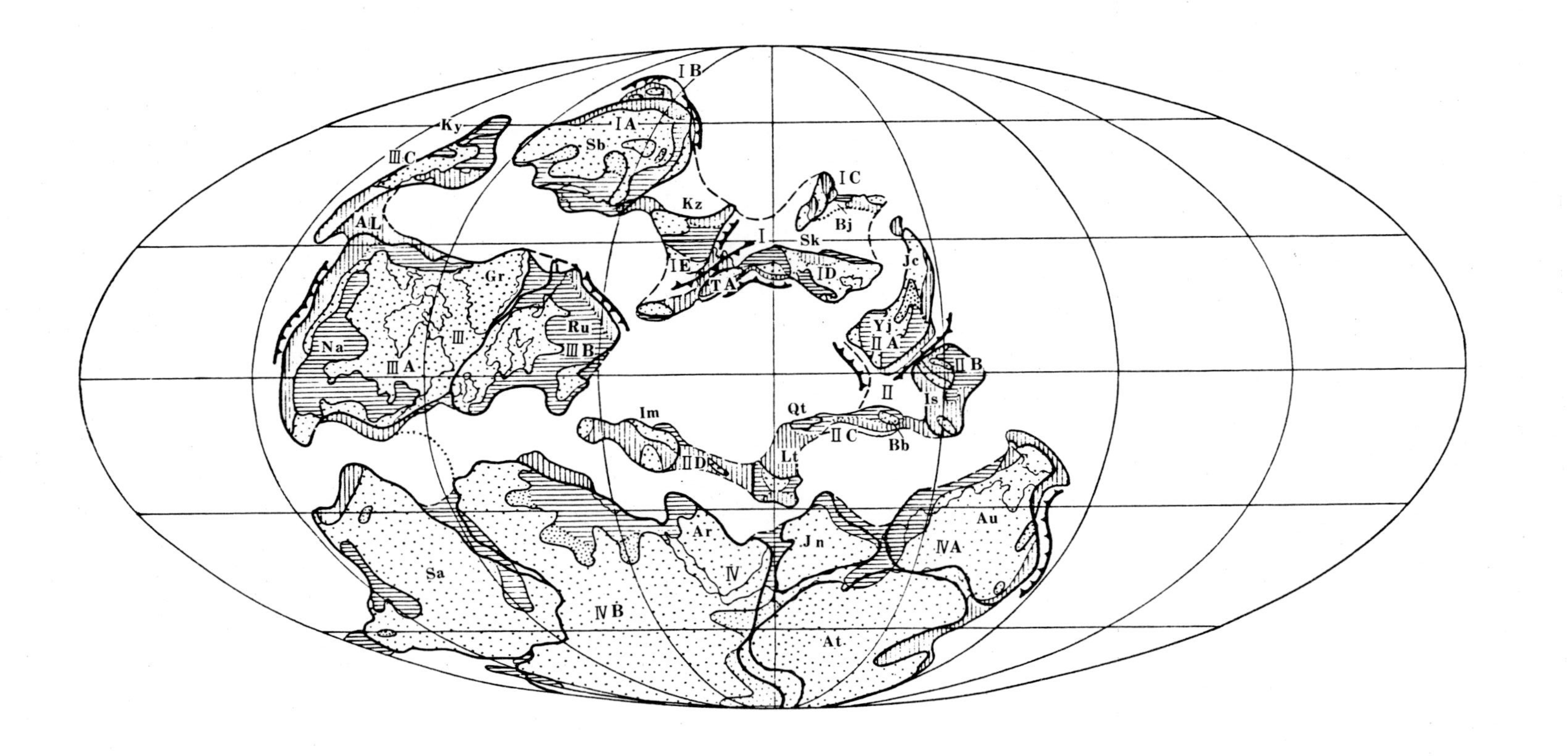

Fig. 3 World reconstruction map showing the Early Carboniferous tectono-palaeogeography

1. Land; 2. Shallow sea deposits; 3. Terrestrial deposits; 4. Deeper sea deposits; 5. Accretional crustal consumption zones; 6. Convergent crustal consumption zones; 7. Tectonic domain boundaries; 8. Geotectonic units. Geotectonic units I NORTH ASIAN (SINO-ANGARAN) DOMAIN IA Siberian subdomain; IB Mongolia-Hinggan subdomain; IC Bureya-Jamusi subdomain; ID Sino-Korean subdomain; IE Kazakhstan-Tarim subdomain. II SOUTH ASIAN DOMAIN IIA Yangtze subdomain; IIB South Sea-Indosinian subdomain; IIC Burma-Tibetan subdomain; IID Ibelia-Iranian subdomain. III EURAMERICA DOMAIN IIIA North American subdomain; IIIB European subdomain. IV GONDWANAN DOMAIN IVA East Gondwanan subdomain; IVB West Gondwanan subdomain.

the world, mainly according to their geotectonic setting. In terms of taxonomy, the Lower Carboniferous Rugosa, apart from the solitary, small, mostly ahermatypic Zaphrentoidida, which amounts to less than 20% of the total, the majority belong the Caninida, which is characterized by an important inovation in skeletal secretion. The skeletal structures and dominated by parallel growth of long needles of fibres which may or may not grouped into fascicles or trabeculae, while the lamellar skeleton composed of calcite flakes so well developed in the Ordovician and Silurian forms, were almost entirely reduced.[10] A simple calculation of Otzuka similarity coefficients, in combination with consideration of the leading genera, has yielded the biogeographical provinces and subprovinces as shown in Fig. 4. Q-type cluster analysis has also been made and has yielded similar results.

In Fig. 4, the realms, provinces and subprovinces and their approximate boundaries are drawn. In addition to the continents and seas, glacigene and evaporite deposits are also shown. It is to be noted that a symmetric distribution of bipolar belts bearing similar faunas in both hemispheres, especially in regard to the brachiopods, is quite pronounced. Indeed, the *Dowhatania-Adminiculoria* fauna in the Himalaya region may, according to Yang Shipu, well be referred to the Gondwanan realm, instead of the South Tethyan province under the Tethyan realm. But as the coral faunas in the Himalaya and the Gangdise regions are similar to each other, and as both are comparable with the Eastern Australian forms, we have preferred to retain the latter regions in the South Tethyan province. Another point to be discussed is the relation between the Yangtze platform and the Qiangtang massif. The occurrence of typical Visean coral and brachiopod faunas of a Yangtzean type,[5] which contain such corals as *Heterocaninia* and brachiopods such as *Datangia* and *Striatifera,* on the southern part of the Qiangtang massif, points strongly that it may be attributed to the Yangtze subprovince, although Late Carboniferous sediments and faunas, the glacigene deposits or diamictites and the cold water eurydesmids, in nearby areas,[11] indicate clearly a Gondwanan affinity. Marine faunas in Jilin and in southwestern Japan was akin to those of Yangtze,[9] and the Jilin-SW Japan subprovince is recognized and attributed to the North Tethyan province. The southern boundary of the Boreal realm, i.e. of the Nei Mongol-Kazakhstan province, needs more investigation. The seaways in the Tianshan and Qilian regions in the Early Carboniferous seem to be quite complicated and transitional type of faunas certainly occurred. In addition, discrepancies and even contradictions in regard to palaeomagnetic data no far obtained in Tarim and North China cannot be easily disposed.[12] But it seems propable that Tarim was not combined to North China in the Late Palaeozoic until in Late Permian. On the contrary, Tarim may have been more akin to Kazakhstan during several periods in the Palaeozoic. In any case, it seems that the old idea of a simple triangular Tethys ocean opening to the east between an integrated Eurasia and a united Gondwana land is not adequate. Instead, island groups affiliated to the main continents and separating oceanic basins of different sizes may be a more reasonable interpretation. In this context, it may be pointed out that how important a role geotectonic thinking may play in biogeographic and palaeogeographic considerations and interpretations.

Table 2 Distribution of Tournaisian rugose coral genera in various parts of the world

	1	2	3	4	5	6	7	8	9	10	11	12	13
Streptelasmatidae													
Triplophyllum					×								
Amplexoididae													
Amplexus	×				×								
Petraiidae													
Thuriantia						×							
Laccophyllidae													
Friedbergia								×					
Barrandeophyllum	×												
Trochophyllum													×
Stereolasmatidae													
Drewerelasma	×			×									
Saleelasma						×							
Guerichophyllidae													
Czarnockia								×					
Hapsiphyllidae													
Allotropiphyllum						×							
Amplexizaphrentis												×	
Clinophyllum	×				×								
Hapsiphyllum					×								
Meniscophyllum					×								
Zaphrentites	×				×	×		×	×	×		×	
Zaphrentoididae													
Dipterophyllum													×
Homalophyllites													×
Menophyllum						×							
Sychnoelasma	×					×		×	×	×	×	×	×
Zaphrentoides	×												
Cyathaxonidae													
Cyathaxonia					×								
Bradiphyllidae													
Fasciculophyllum					×								
Soshkineophyllum					×								
Plerophyllidae													
Dalnia								×					
Palaeosimilia						×	×	×	×	×	×	×	
Polycoelidae													
Calophyllum					×								
Kinkaidia					×								
Adamanophyllidae													
Cravenia	×												
Tachyphyllum												×	
Lophophyllididae													
Lophyllidium					×								
Caninidae													
Caninia	×	×	×	×		×	×			×	×	×	
Ceriphyllum	×												

(to be continue)

	1	2	3	4	5	6	7	8	9	10	11	12	13
Complanophyllum	×												
Bothrophyllidae													
Bothrophyllum			×							×			
Campophyllum						×							
Caninophyllum	×		×										
Kusbassophyllum			×										
Uralinidae													
Amunikephyllum				×									
Bifossularia												×	
Beichuanophyllum		×	×	×									
Cystophrentis	×	×					×		×		×		
Humboldtia	×		×	×									
Kakwiphyllum				×									
Keyselingophyllum	×					×	×		×				
Pseudouralina	×	×	×										
Uralina	×	×					×	×	X		×	×	
Vesiculophyllum	×												
Siphonophyllia	×	×	×										
Palaeosimilidae													
Molophyllum											×		
Koninckophyllidae													
Arachnolasma												×	
Eostrotion						×							
Koninckophyllum			×										
Lophphyllum						×							
Amygdalophyllidae													
Amygdalophyllum	×			×									
Corruthersella	×												
Ekvasophyllum	×												
Clisiophyllidae													
Clisiophyllum			×										
Cyathoclisia						×	×	×	×			×	
Dibunophyllidae													
Dibunophyllum			×	×									
Lithostrotionidae													
Diphyphyllum	×												
Donophyllum				×									
Kwangsiphyllum		×	×										
Lithostrotion	×												
Aulinidae													
Aulostylus													×
Solenodendron						×			×				
Vesiculotubus												×	
Thysanophyllidae													
Dematophyllum	×		×										
Stelechophyllum	×	×	×									×	
Thysanophyllum	×												

1. Yangtze; 2. Longmenshan; 3. South Qinling; 4. Qaidam-South Qilian; 5. North Tianshan; 6. Northwest Europe; 7. Central and Southern Europe; 8. East Europe; 9. Ural; 10. Central Asia; 11. Far East; 12. Kuznetzk; 13. Eastern North America.

Table 3 Distribution of Visean rugose coral genera in various parts of the world

	1	2	3	4	5	6	7	8	9	10	11	12	13	14	15	16	17
Amplexoididae																	
Amplexus				×													
Spongophyllidae																	
Spongophyllum							×										
Stringophyllidae																	
Parasociophyllum			×														
Sunophyllidae																	
Acmophyllum	×																
Eridophyllidae																	
Eridophyllum	×																
Petraiidae																	
Cystelasma																	×
Lacophyllidae																	
Barrandeophyllum					×			×									
Trochophyllum																	×
Neaxon									×								
Metriophyllidae																	
Metriophyllum																×	
Stereolasmatidae																	
Saleelasma									×								
Hapsiphyllidae																	
Allotropiophyllum	×								×							×	
Amplexozaphrentis	×						×		×	×	×						×
Barytichisma				×												×	
Canadiphyllum																×	
Hapsiphyllum				×	×	×										×	×
Longicalava																×	
Meniscophyllum								×								×	
Neozaphrentis																×	
Zaphrentites	×							×	×		×	×	×				
Zaphrentoididae																	
Ankhelasma																×	
Commusia																	×
Dipterophyllum																	×
Homalophyllites																×	×
Sychnelasma												×					
Zaphrentoides									×								
Cyathaxonidae																	
Cyathaxonia	×			×	×			×	×	×		×			×	×	
Bradyphyllidae																	
Claviphyllum									×								
Rotiphyllum	×			×	×			×	×				×				
Soshkineophyllum								×									

(to be continue)

	1	2	3	4	5	6	7	8	9	10	11	12	13	14	15	16	17
Plerophyllidae																	
Amplexocarina																×	
Baryphyllum																	×
Pentaphyllum									×								
Tachylasma	×				×			×									
Ufimia									×		×						
Polycoelidae																	
Calophyllum									×								
Kinkaidia																	×
Tetralasma							×		×								
Adamanophyllidae																	
Adamanophyllum	×							×									
Cravenia									×			×					
Flagellophyllum								×									
Lophophyllididae																	
Lophamplexus																	×
Lophophyllidium				×													
Stereostylus	×																
Sugiyamaella				×			×					×					
Caninidae																	
Allotabulophyllum	×																
Arctophyllum												×					
Caninia	×		×	×	×		×		×	×			×			×	
Corphalia									×								
Fomitchevella							×					×					
Haplolasma									×	×	×	×				×	×
Lublinophyllum											×						
Melanophyllum	×	×	×									×					
Pseudozaphrentoides									×	×			×	×		×	×
Bothrophyllidae																	
Bothrophyllum									×								
Calmissiphyllum	×								×		×						×
Campophyllum	×								×		×						×
Caniniella												×					
Caninophyllum	×			×					×	×					×		
Heterocarinia	×		×		×		×	×					×				
Kueichouphyllum	×		×	×			×	×		×	×	×	×				
Kusbassophyllum	×	×															
Uralinidae																	
Keyselingophyllum	×								×	×							
Liardiphyllum																×	
Merlewoodia															×		
Siphonophyllia	×	×		×			×	×	×	×	×		×				
Palaeosimilidae																	
Katranophyllum													×				
Palaeosimilia	×	×	×	×		×	×	×	×	×	×	×	×	×			
Palaeostraea	×	×		×			×	×	×	×		×	×				×

(to be continue)

	1	2	3	4	5	6	7	8	9	10	11	12	13	14	15	16	17
Qinghaiphyllum	×		×	×													
Symplectophyllum															×		
Koninckophyllidae																	
Arachnolasma	×	×	×	×	×		×	×		×	×		×	×			
Heintzella											×	×					
Koninckophyllum	×				×		×		×	×	×	×	×				×
Neokoninckophyllum											×						
Yuanophyllum	×	×	×		×	×	×	×					×	×			
Zakowia											×						
Amygdalophyllidae																	
Amygdalophyllum	×			×					×	×	×	×		×	×		
Arachnolasmella												×	×				
Carruthersella	×		×	×					×	×	×						
Ekvasophyllum	×															×	
Kazachiphyllum												×	×				
Ramiphyllum	×																
Rozkowskia											×				×		
Rylstonia	×				×		×		×		×						
Pseudopavonidae																	
Cionodendron															×		
Spriophyllum			×						×		×	×					
Clisiophyllidae																	
Caninostrotion	×				×				×								×
Clisiophyllum	×			×		×	×	×	×	×	×		×	×			
Cyathoclisia									×	×	×	×					
Cysticlisiophyllum				×													
Hunanoclisia	×		×			×											
Neoclisiophyllum	×		×		×	×	×	×	×	×	×	×	×				
Dibunophyllidae																	
Corwenia	×		×	×			×	×	×		×	×	×				
Cystikoninckophyllum	×																
Debaophyllum	×																
Dibunophyllum	×	×	×	×	×	×	×	×	×	×	×			×		×	×
Nagatophyllum														×			
Aulophyllidae																	
Auloclisia	×				×	×			×	×		×					
Aulophyllum				×					×	×	×			×			
Axoclisia										×							
Biphyllum											×						
Nervophyllum	×																
Staurophyllum												×					
Lithostrotionidae																	
Akiyosiphyllum														×			
Arachnastraea	×						×										
Aulostrtion														×			

(to be continue)

	1	2	3	4	5	6	7	8	9	10	11	12	13	14	15	16	17
Diphyphyllum	×		×	×		×	×	×	×	×	×	×	×	×	×	×	
Donophyllum	×	×	×														
Kwangsiphyllum	×						×										
Nemistium									×		×	×	×				
Lithostrotion	×	×	×	×	×	×	×	×	×	×	×		×	×	×	×	
Siphonodendron	×			×	×	×	×		×	×	×	×	×	×	×	×	×
Aulinidae																	
Aulina	×	×		×	×		×	×	×		×	×		×		×	
Aulokoninckophyllum			×	×	×				×	×	×	×			×		
Aulostylus																×	
Paraaulina	×																
Solenodendron									×			×					
Veseaulina									×								
Vesiculotubus	×			×													
Durhaminidae																	
Protodurhamina											×						
Aphrophyllidae											×						
Aphrophyllum	×	×	×												×		
Aphrophylloides															×		
Kizilia	×	×				×						×	×				
Melanophyllidium													×				
Nothaphrophyllum															×		
Thysanophyllidae																	
Acrocyathus											×					×	×
Dorlodotia	×	×		×					×			×			×		
Sciophyllum														×		×	
Stelechophyllum	×										×	×				×	
Thysanophyllum	×			×					×		×	×				×	
Petalaxidae																	
Paralithostrotion	×							×			×	×	×				
Petalaxis	×	×		×		×		×									
Lonsdaleidae																	
Actinocyathus	×	×		×	×		×	×	×					×			
Axophyllum									×	×			×	×	×		
Gangamophyllum	×		×	×	×	×	×	×		×	×	×	×	×			
Lonsdaleia	×	×	×	×	×	×	×	×	×		×	×		×			×
Pareynia										×							
Kionophyllidae																	
Parakoninckophyllum	×																
Waagenophyllidae																	
Huishuiphyllum	×																
Wentzellophyllidae																	
Majiaobaphyllum	×	×	×														

1. Yangtze; 2. Longmenshan; 3. South Qinling; 4. Qaidam-South Qilian; 5. North Qilian; 6. Sanjiang-Qamdo; 7. North China; 8. North Tianshan; 9. Northwest Europe; 10. Central and Southern Europe; 11. East Europe; 12. Ural; 13. Central Asia; 14. Japan; 15. Eastern Australia; 16. Western North America; 17. Eastern North America.

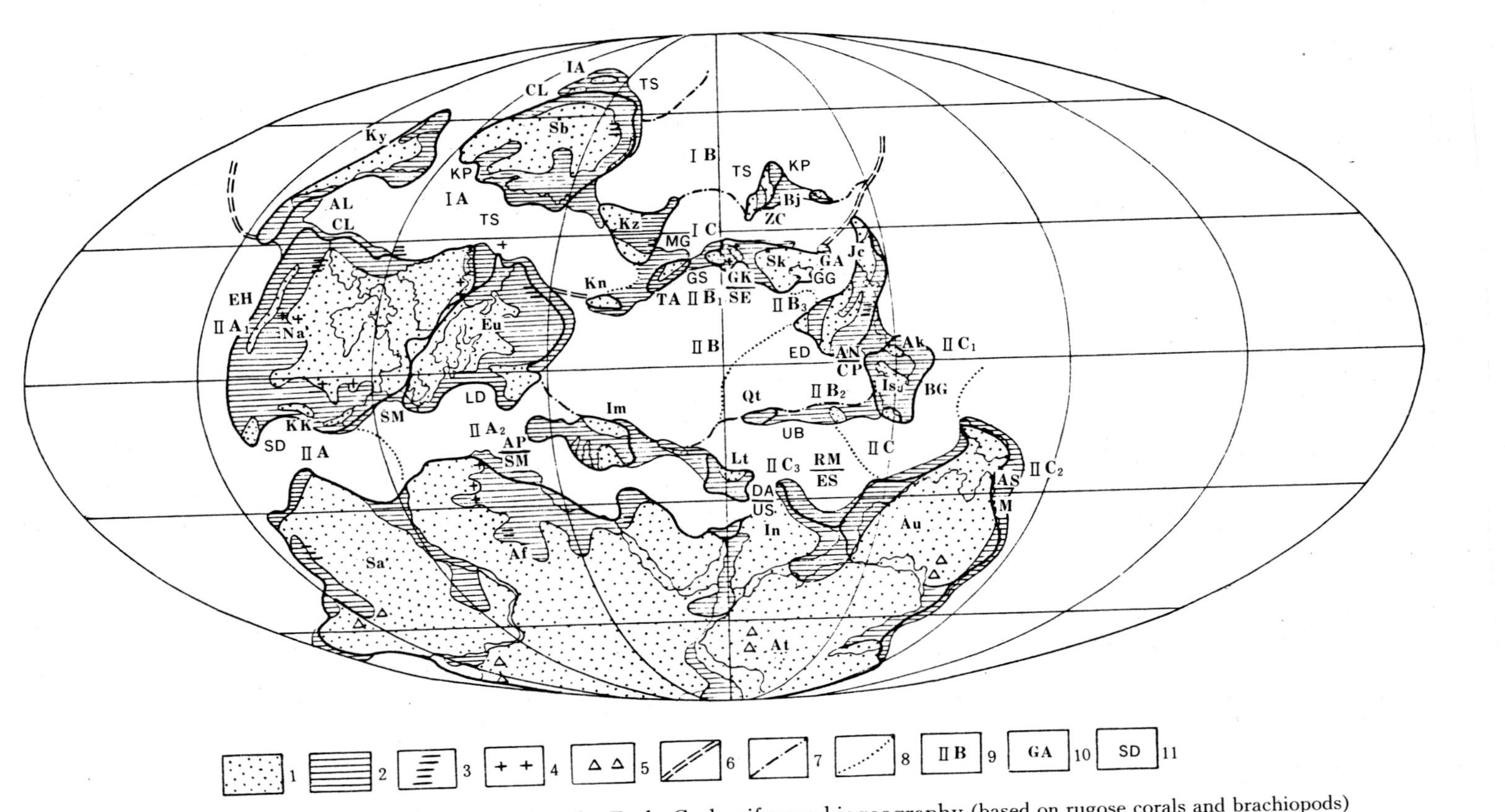

Fig. 4 World reconstruction map showing the Early Carboniferous biogeography (based on rugose corals and brachiopods)

1. Land; 2. Marine and terrestrial deposits; 3. Coal deposits; 4. Evaporite deposits; 5. Glacigene deposits; 6. Boundary between biogeographic realms; 7. Boundary between biogeographic provinces; 8. Boundary between biogeographic subprovinces; 9. Biogeographic units; 10. Rugose coral assemblages; 11. Brachiopod assemblages.

I BOREAL REALM IA Siberian province Co: CL *Canadiphyllum-Liardiphyllum*; Br: ST *Setigerites-Tomilia*. IB Mongolia-Hinggan province Co: RU *Rotiphyllum-Uralinia*; Br: KP *Kitakamithyris-Prehorridonia*, TS *Tolmatchoffia-Syringothyris*. IC Nei Mongol-Kazakhstan province Co: GK *Gangamophyllum-Kueichouphyllum*, ZC *Zaphrentites-Cyathaxonia*; Br: MG *Marginatia-Gigantoproductus*. II TETHYAN REALM IIA West Tethyan province IIA1 North American subprovince Co: KK *Kakwiphyllum-Kinkaidia*, EH *Ekvasophyllum-Homophyllites*; Br: SD *Stegacanthia-Diaphragma*. IIA2 European subprovince Co: AP *Axolinia-Pareynia*, SM *Siphonophyllia-Menophyllum*; Br: LD *Levitusia-Delipinea*. IIB North Tethyan province IIB1 Tarim-Qinling subprovince Co: GK *Gangamophyllum-Kueichouphyllum*, SE *Siphonophyllia-Enygmophyllum*; Br: GS *Gigantoproductus-Striatifera*, MS *Marginatia-Syringothyris*. IIB2 Yangtzean subprovince Co: AH *Arachnolasma-Heterocaninia*, CP *Cystophrentis-Pseudouralinia*; Br: ED *Eochoristites-Datangia*. IIB3 Jilin-SW Japan subprovince Co: GA *Gangamophyllum-Amygdalophyllum*, ZC *Zaphrentites-Cyathaxonia*; Br: GG *Gigantoproductus-Grandispirifer*. IIC South Tethyan province IIC1 Southeast Asian subprovince Co: AK *Amygdalophyllum-Kueichouphyllum*, Br: BG *Balakhonia-Gigantoproductus*. IIC2 East Australian subprovince Co: AS *Amydalophyllum-Symplectophyllum*, M *Merlewoodia*. IIC3 South Tibet (Xizang) subprovince Co: RM *Rhopalolasma-Mirusophyllum*, ES *Ekvasophylloides-Sychnoelasma*; Br: DA *Dowhatania-Adminiculoria*, US *Unispirifer-Syringothyris*.

DISTRIBUTION OF SOME IMPORTANT CARBONIFEROUS SEDIMENTARY MINERAL DEPOSITS IN CHINA

Carboniferous strata are widespread in China. Apart from the most important coal deposits, there are also gypsum, bauxite and different types of iron deposits, including the well-known Shanxi type partly of residual origin, and the volcano-sedimentary type of even more economic importance. The distribution of these mineral deposits is mainly controlled by climatic and palaeogeographic and tectonic conditions, as may be clearly shown in the maps (Figs. 5, 6).

Gypsum deposits of Early Carboniferous age are widespread in Northwest China. The gypsum-bearing beds are widely distributed along the northwestern border of the Tarim platform, occurring in the Karashai and the Bachu Formations. In the Qilian Mountains and the Gansu Corridor region, between the Qaidam massif in the south and the Alxa massif in the north. To the north of the Longxi massif, Lower Carboniferous littoral to lagoonal sediments, the Qianheishan Formation and the Chouniugou Formation, contain rather big gypsum deposits, occasionally also rock salts. Similar sedimentary deposits are known also in the southern Qilian, in the Huaitoutala Formation and the Chengqianggou Formation.

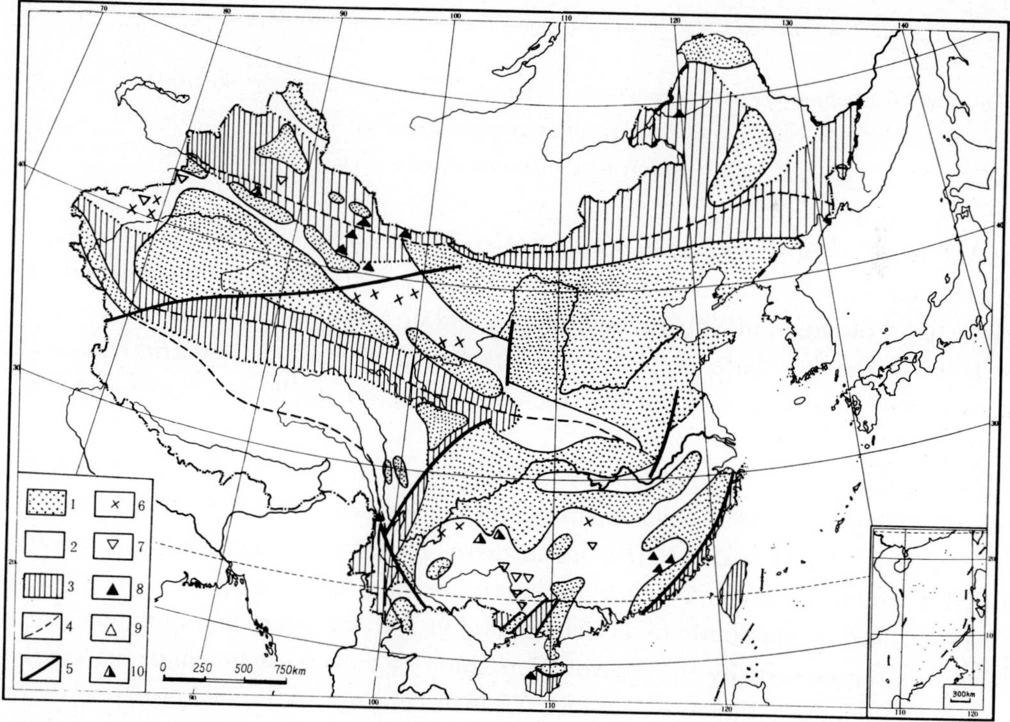

Fig. 5 Early Carboniferous palaeogeographic map of China showing the distribution of the main sedimentary mineral deposits

1. Land; 2. Shallow sea; 3. Deeper sea; 4. Subsequent convergent crustal consumption zones; 5. Subsequent fault zones; 6. Gypsum deposits; 7. Manganese deposits; 8. Iron deposits; 9. Bauxite deposits; 10. Iron and bauxite deposits.

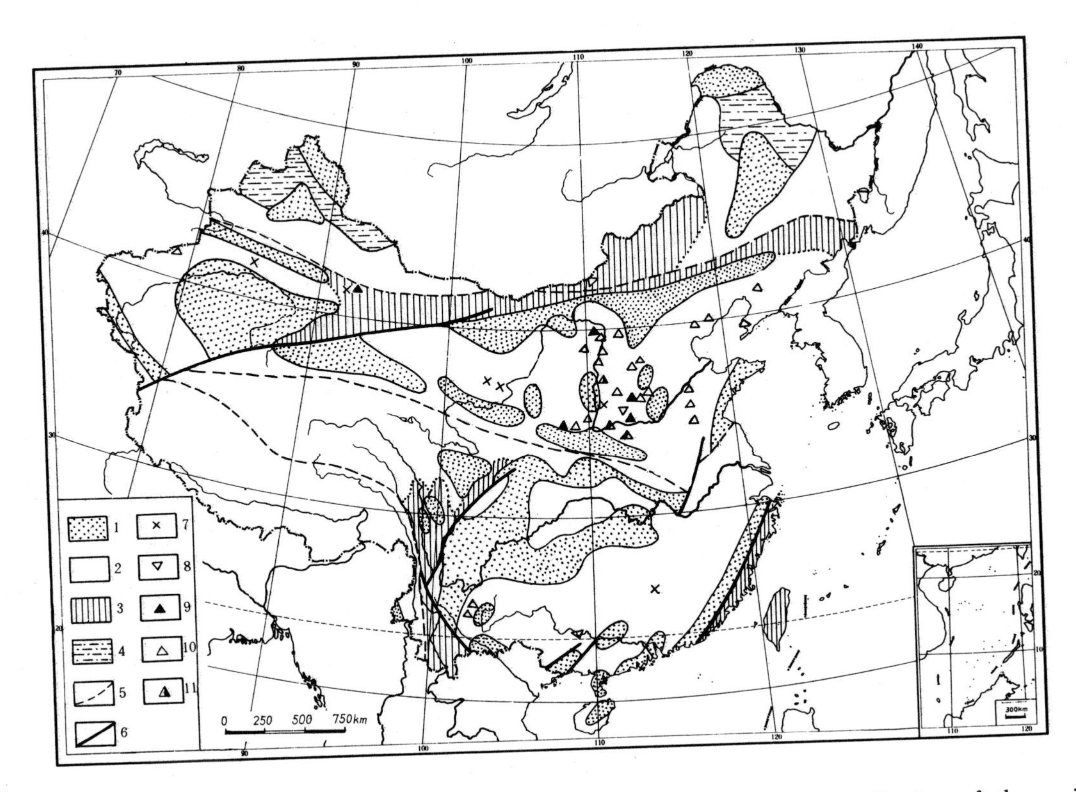

Fig. 6 Late Carboniferous palaeogeographic map of China showing the distribution of the main sedimentary mineral deposits

1-3 as in Fig. 5; 4. Mainly terrestrial basin deposits; 5-11 as in 4-10 of Fig. 5.

Another belt of evaporite deposits is distributed along the southern border of the Yangtze platform, among which the best known are the Visean gypsum beds in the Zimengqiao Formation of Shuangfeng, Hunan and of Yongxin, Jiangxi.

It is evident that these evaporite deposits owe their origin to the arid climatic zones where extensive, partly restricted epeiric seas occur on the margin of the stable platforms.

Lower Carboniferous iron deposits are of two types and are usually associated with manganese deposits. The geosynclinal type of volcano-sedimentary origin in the northern belt is important and is distributed in northern Tianshan, in the Visean Yamansu and Akshak Formations, which also contain important manganese deposits. It was also found in Beishan and in the southern Da Hinggan Mts., in the Xiertala Formation. In South China, it is represented by the Mazara Formation of volcano-sedimentary and carbonate rocks in the Sanjiang Valley of western Yunnan, and by the Lindi Formation of similar lithology in western Fujian.[4] It may be noted that, the famous Shilu iron ore of Hainan Island, hitherto attributed to Carboniferous age, may, according to recent investigations, turn out to be of much older age. Visean manganese deposits associated with carbonates and silicolites, which abound in many parts of eastern Guangxi, notably in Yishan and

Nandan, are probably of open shelf type in the semistable sea to the south of the Yangtze platform. Important bauxite deposits of Visean age are known in Guizhou in the upper Yangtze, which are closely associated with contemporary iron deposits.

In the Late Carboniferous, the palaeogeographical outline of China bears no much difference from the Early Carboniferous, except for the extensive marine inundation in North China (Figs. 5, 6).[13]

Late Carboniferous evaporites containing gypsum deposits are often present in association with Early Carboniferous ones in Northwest China. They occur in the Karamiran Group near Qiemo, southern Tarim, in the Tugubulak Formation near Hami, Xinjiang, and in the Yanghukou Formation in the Gansu Corridor. Iron and manganese deposits of Late Carboniferous age are known in the Huanglung Formation of the lower Yangtze region.

By far the most important are the Late Carboniferous platform type bauxite deposits of North China, which are closely associated with the residual type of Shanxi iron deposits. The carboniferous bauxite deposits are the most important in the aluminum reserve of China, and the Chinese Carboniferous bauxites are also the only important bauxite deposits of Carboniferous age ever known in the world.

The so-called Shanxi type of iron and aluminum deposits are extensively distributed in North China, mostly transgressive over the erosional surface of the karst topography formed of Lower Palaeozoic, mainly Lower Ordovician limestones, and thus different in horizon in different places. The palaeogeographic outline in North China in the Late Carboniferous has been made rather clear.[13] The northern part of North China was flooded by the Late Carboniferous Penchian (Moscovian) epicontinental seas which came both from Liaoning in the east, and Gansu Corridor in the west. This transgression reached as far south as the border line between Shanxi and Hanan in the west at about 35°N latitude and between Shandong and Jiangsu in the east at 34°30′N latitude. Most parts of North China to the south of this line had lain above sea level in the Moscovian age. Another seaway came from the east, from northern Jiangsu via Xuzhou. In the western part of North China, a fault zone bounded the west border of the Ordos massif, to the west of which the Upper Carboniferous becomes suddenly coarser and was no less than ten times thicker than in the east. The souther part of North China began to be transgressed by the Qinling sea from the south, not until in the Uralian or Taiyuanian time, and also in the Early Permian. This northerly transgressing sea reached approximately the same latitude as the Moscovian marine ingression did from the opposite direction. Further to the north was developed the extensive swampy lands which gave rise to the very important coal deposits of Taiyuanian and Shanxian age. The sedimentary sequence of the iron and bauxite deposits consists in general of the siderite beds below and the bauxites beds above. The ideal environment for bauxite accumulation seems, according to Meng Xianghua et al.,[14] to be in current channels and tidal flats on the restricted low energy coastal areas bounded by low uplift in the extensive littoral belt.

REFERENCES

1. Wang Hongzhen, Earth Scien—Jour Wuhan Coll Geol, 1 (1981), 42.
2. ——, In: Tectonic History of the Ancient Continental Margin of South China, Wuhan College of Geology Press (1986), 1.
3. ——, In: The Geology of China, Oxford Monographs on Geology and Geophysics 3, Oxford University Press (1986), 235.
4. Luo Jinding and Yan Qingnan, Sci Geol Sinica, (4) (1984), 332.
5. Yang Shipu and Fan Yingnian, Contribution to the Geology of the Qinghai-Xizang(Tibet) Plateau, 10 (1982), 65.
6. Luo Jinding, Palaeontographica Americana, 54 (1984), 427.
7. Rowley DB, Raymond A, Parrish JT et al, Interat Jour Coal Geol, 5 (1) (1985), 7.
8. Lin Jinlu, Seismology and Geology, 9 (2) (1987), 91.
9. Minato M and Kato M, 9^{e} Congr Int Strt Geol Carb Washington and Champaign-Urbana, 1979, C R 3 (1984), 256.
10. Wang Hongzhen and He Yuanxiang, Acta Palaeont Sinica, 24 (2) (1985), 134.
11. Liu Benpei and Cui Xinsheng, Earth Science — Jour Wuhan Coll Geol, 1 (1983), 79.
12. Bai Yunhong, Chen Guoliang et al, Tectonophsics, 139 (1987), 145.
13. Wang Hongzhen(chief compiler), In: Atlas of the Palaeogeography of China, Cartographic Publishing House (1985), 69.
14. Meng Xianghua, Ge Ming and Xiao Zengqi, Acta Geol Sinica, 67 (2) (1987), 182.

XIe Congrès International de Stratigraphie et de Géologie du Carbonifère Beijing 1987,
Compte Rendu 1 (1991): 117-141

THE EXPLORATION AND EXPLOITATION OF COAL RESOURCES IN CHINA

Wang Zhongtang
(Geological Bureau, Ministry of Coal Industry of China)

INTRODUCTION

During 38 years since the founding of the People's Republic of China, we have won great achievements in the production and construction of coal industry. The output of raw coal had increased from 32.43 million tons in 1949 to 894.04 million tons in 1986. Among the 2 313 counties and cities in our country, 1 458 are possessed of coal resources, accounting for about one eighteenth of our total territory. Until the end of 1986, the demonstrated coal reserves are 8 597 hundred million tons and the available demonstrated reserves are 8 458 hundred million tons. The bituminous coal makes up 59% of total reserves, while the anthracite 13%, the brown coal 14.6%, and the others 13.4% respectively. The coal resources are most guaranteed as compared with those of other mineral resources in our country.

The coal consumption used to be always more than 70% of total energy consumption. In 1985, for example, coal consumption is 72.8%, petroleum 20.9%, natural gas 2%, while hydroelectricity 4.3%, and this composition of energy will be retained till the end of this century . Looking forward to the future, the coal industry still needs to be developed actively, especially for those old coal industry bases in eastern China and those new bases in central and western China, where the exploration and exploitation of coal resources are very urgent.

FORMATION AND DISTRIBUTION OF COAL RESOURCES IN CHINA

1. Main coal-forming period

China is one of few countries which have the richest coal resources in the world, with characteristic of multi-coal-forming periods, wide dispersion, a complete range of coal ranks and better conditions for exploitation(Fig. 1). The stone-like coal (a kind of sapropelic coal) of the Early Palaeozoic Era is the oldest coal in China. From then to the Quaternary, there are fourteen coal-forming periods. The earliest coal having important economic value was formed during the Early Carboniferous Tatangian age. The most important coal-forming period in our country is the Permo-Carboniferous in North China,

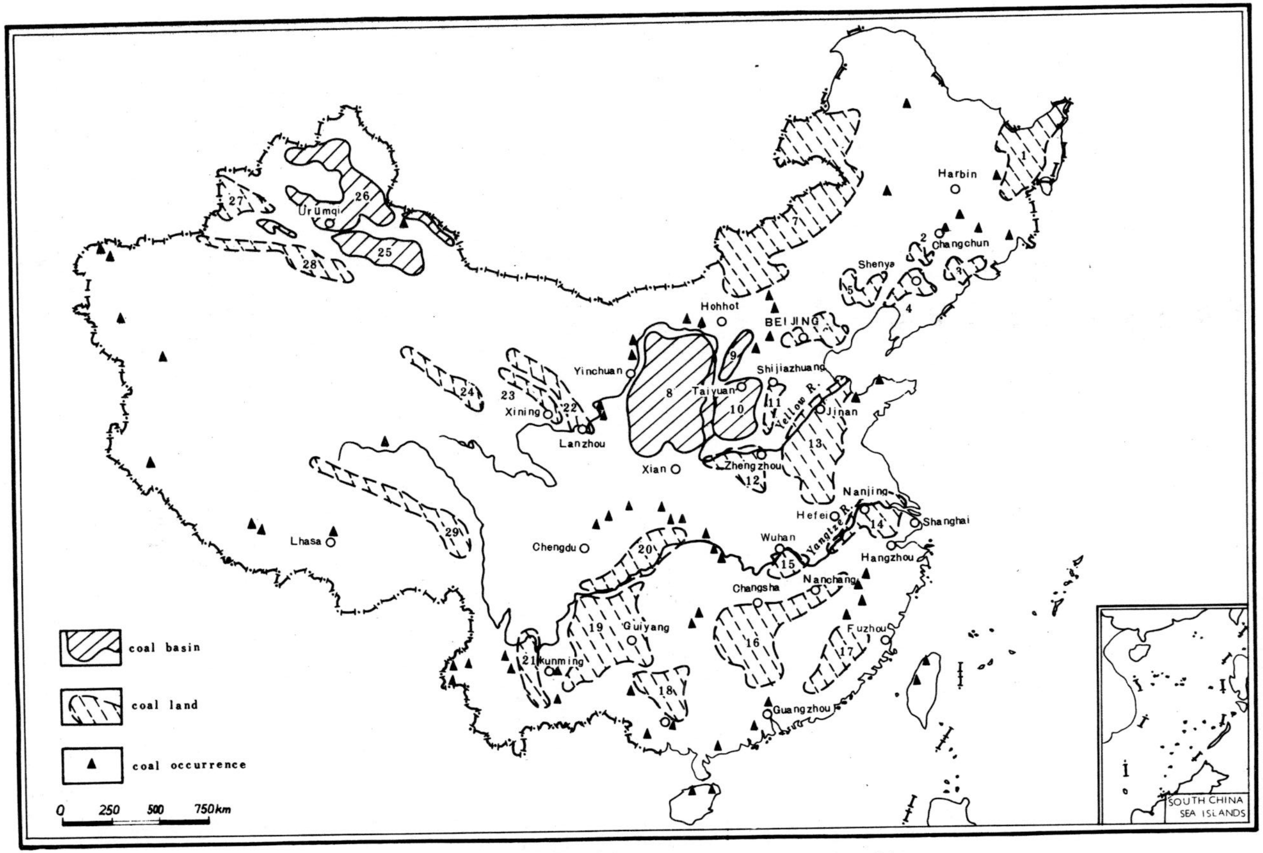

Fig. 1 Sketch map of coalfield distribution of China

1. Sanjiang-Muling area; 2. North Liaoning area; 3. Hun jiang area; 4. Liaohe-Taizihe area; 5. West Liaoning area; 6. Jingtang area; 7. Eastern Nei Mongol; 8. Coal basin of Ordos; 9. Coal basin of Daning; 10 Coal basin of Qinshui; 11. Eastern Taihang; 12. West Henan area; 13. Jiangsu-Shandong-Henan-Anhui area; 14. Southern Zhejiang-Jiangsu Anhui; 15. Southeast area of Hubei; 16. Huang-Jiangxi-Gungdong area; 17. Fujian-Guangdong area; 18. Central Guangxi area; 19. Guizhou-Yunnan-Sichuan area; 20. Huayingshan area; 21. Dianzhong area; 22. Hexi Corridor area; 23. Datonghe area; 24. Northern Qaidam area; 25. Coal basin of Turpan-Hami; 26. Coal basin of Junggar; 27. Yili areea; 28. Northern Tarim; 29. Northern Xizang area.

the Permian and the Late Triassic in South China, the Early and Middle Jurassic in Northwest China, the Late Jurassic-Early Cretaceous in Northeast China and the Tertiary in Northeast and Southwest China and several coastal provinces(Fig. 2).[1]

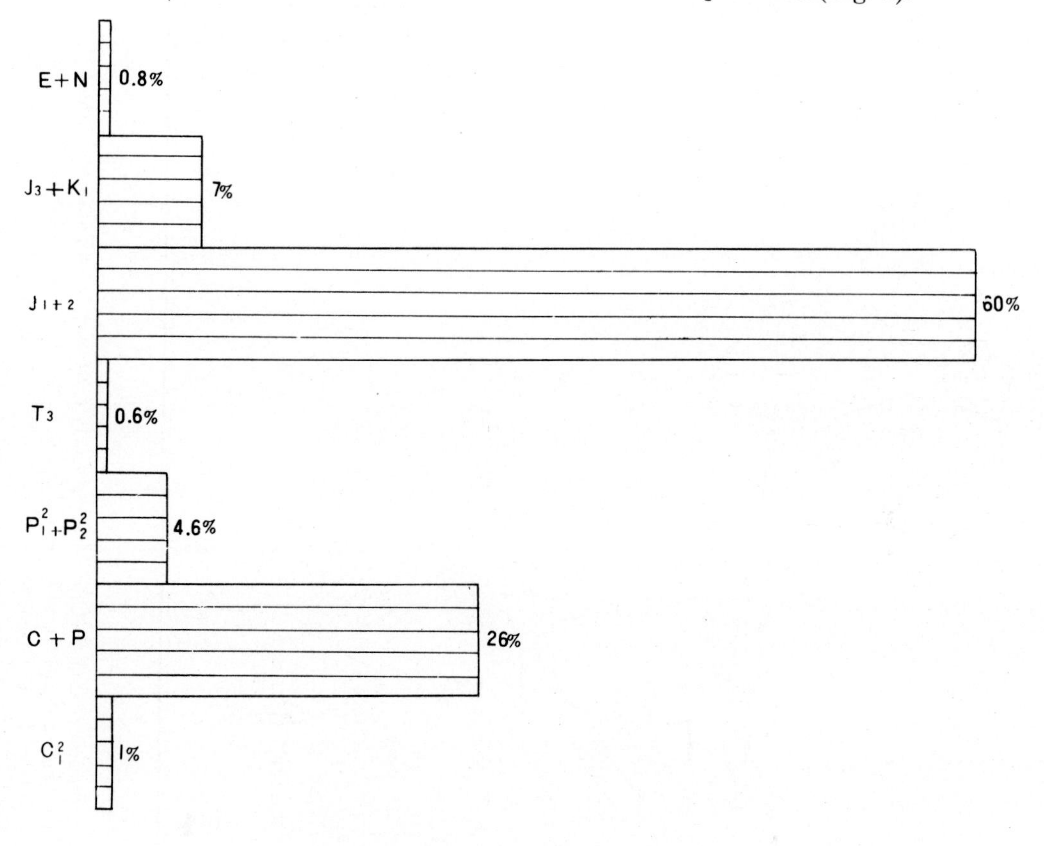

Fig. 2 Constitution diagram of Chinese coal based on their age

Carboniferous is the most important coal-forming period in the world. The Early Carboniferous Tatangian is distributed widely in South and Southwest China, such as in eastern Yunnan, central Hunan, southern Jiangxi and northern Guangdong. During this stage coal used to be formed in the lagoon and bay environments nearby an oldland. At the margin of old upwarped districts, such as Qamdo of Xizang and Shangcheng of Henan, there are several coalfields with thin coal beds deposited in the transitional environment between sea and land.

Coal resources of the Early Carboniferous in regions mentioned above make up about 1% of total coal resources of the country(Fig. 3).

In North China and the eastern part of Northwest China, there was a continuous coal-forming period from the Late Carboniferous to Permian. The coal accumulation center shifted gradually from north to south as time went by. The Penchi Formation (Westphalian) contains thin coal beds to the north of 38^0N latitude, and in some areas, the beds have economical value. In the Taiyuan Formation (Stephanian), coal beds which have economical value are distributed to the north of 34.5^0N latitude(Fig.4).

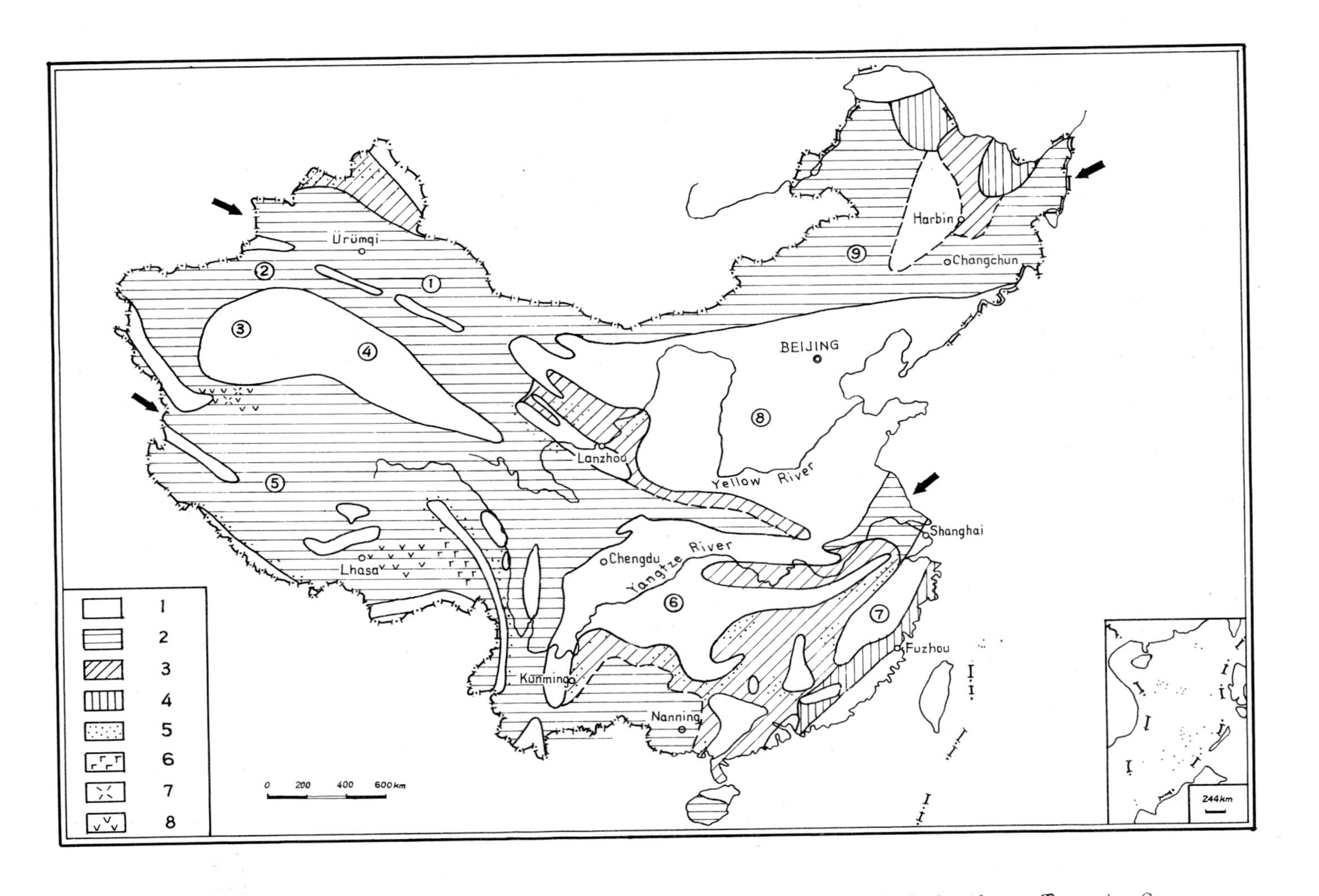

Fig. 3 Palaeogeographic map showing accumulation of coal in China during Early Carboniferous Tatangian Stage
1. Oldland or stripped zone; 2. Marine facies; 3. Transitional facies (Coal-forming deposits); 4. Terrestrial facies; 5. Coal rich zone; 6. Basic volcanic rocks 7. Acidic volcanic rocks; 8. Intermediate volcanic rocks. ① Junggar sea trough; ② Southern Tianshan sea trough; ③ Tarim oldland; ④ Qaidam oldland; ⑤ Kan-Zang sea trough; ⑥ South China upland; ⑦ Zhe-Min oldland; ⑧ North China peneplain; ⑨ Hinggan-Mongolia sea trough.

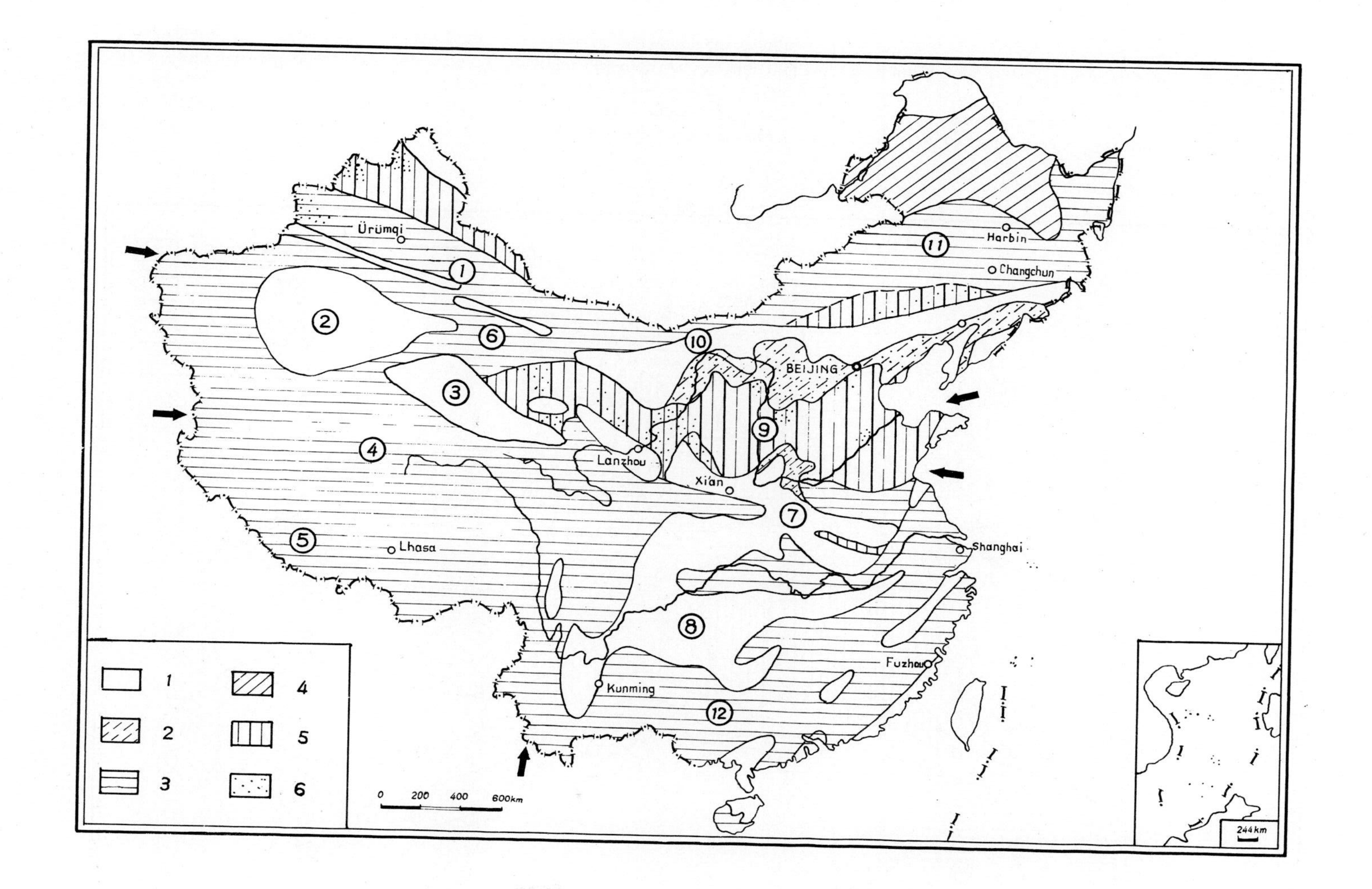

Fig. 4 Palaeogeographic map showing accumulation of coal in China during Middle-Late Carboniferous

1. Oldland or stipped zone; 2. Transitional facies (mainly terrestrial); 3. Marine facies; 4. Alternating marine and terrestrial facies; 5. Alternating marine and terrestrial facies with coal-bearing deposits; 6. Coal-rich zone. ① Junggar sea trough; ② Tarim oldland; ③ Qaidam oldland; ④ Kan-Zang sea trough; ⑤ Himalaya sea; ⑥ Qilian shallow sea; ⑦ Qinling oldland; ⑧ South China oldland; ⑨ North China epicontinental sea; ⑩ Yinshan oldland; ⑪ Hinggan-Mongolia sea trough.

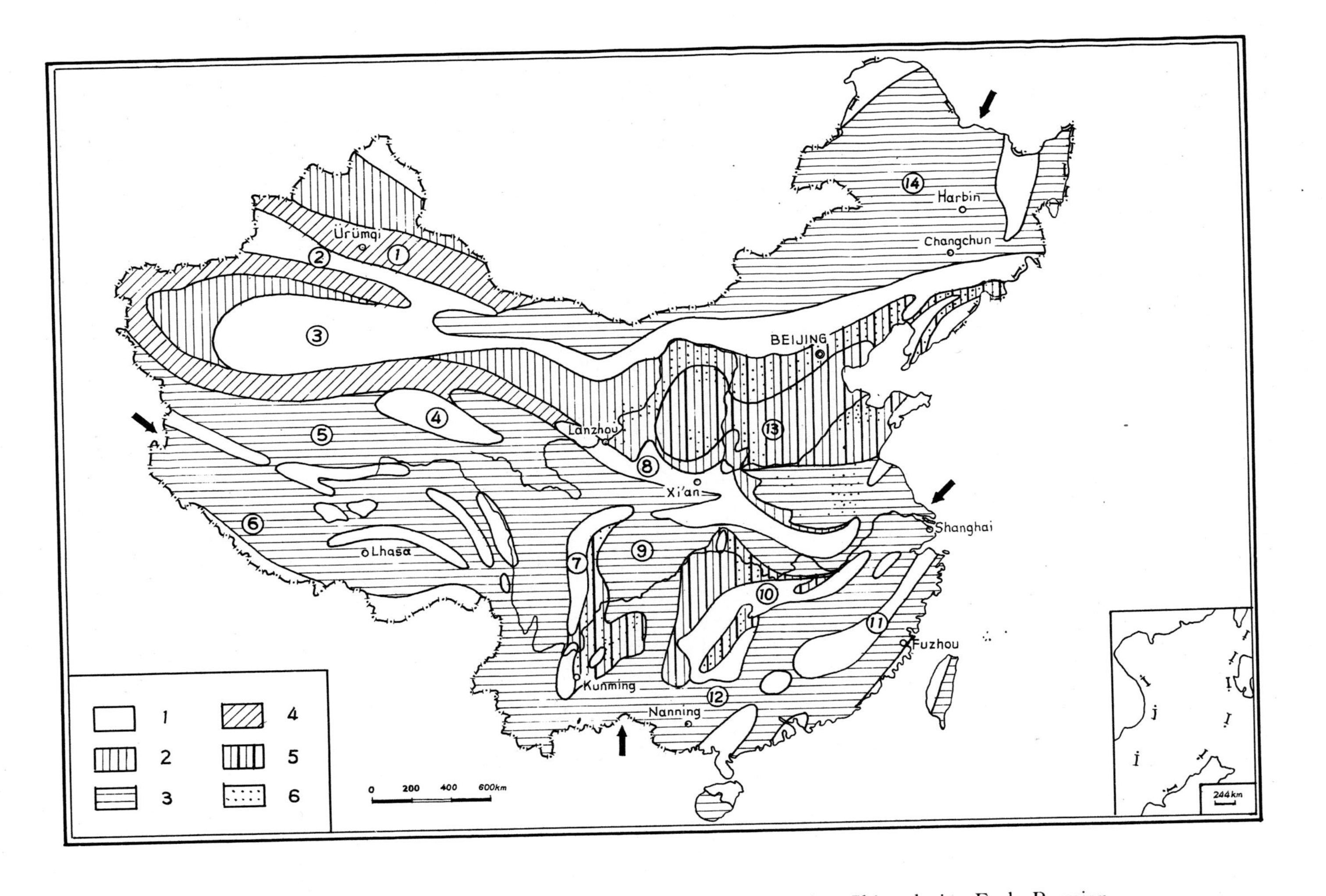

Fig. 5 Palaeogeographic map showing accumulation of coal in China during Early Permian

1. Oldland or stripped zone; 2. Terrestrial facies; 3. Marine facies; 4. Alternating and terrestrial facies; 5. Alternating marine and terrestrial facies with coal-bearing deposits; 6. Coal-rich zone. ① Junggar lowland; ② Tianshan-Yinshan oldland; ③ Tarim oldland; ④ Qaidam oldland; ⑤ Qing-Zang sea; ⑥ Himalaya sea trough; ⑦ Kang-Dian oldland; ⑧ Qin-Huai oldland; ⑨ Yangtze sea; ⑩ Jiangnan oldland; ⑪ Zhe-Min oldland; ⑫ South China sea; ⑬ North China epicontinental sea; ⑭ Hinggan-Mongolia sea trough.

In the Shansi Formation (Lower Rotliegende), from 39^0N latitude to the south, from western Henan to northern Anhui there are many coal beds with high quality(Fig. 5).

In the Shihhotse Formation, the exploitable coal beds are distributed in a limited zone from Xuzhou, Huainan and Huaibei to western Henan in southern North China. The Permo-Carboniferous coal-bearing deposits above-mentioned have reflected a panorama of depositional environments from seashore to delta. Coal reserves of this period make up 26% of total coal resources in China. About 60% output of coal is from Permo-Carboniferous coalfields in this area.

In South China, the Liangshan Formation of the early Early Permian contains several thin coal beds, some of which have economical value. But in South China, the important coal-forming period was from the Maokouan age (later Early Permain) to Changhsingian age (later Late Permian); it was also a continuous coal-accumulating process. The coal-accumulating center of Longyan Formation of Maokouan age of the eastern coastal area, such as southwest Fujian, eastern Jiangxi and northern Zhejiang shifted toward northwest, in pace with the changing paleogeographical environment.

In the Lungtan Formation of the early Late Permian in Hunan, Jiangxi, Guangdong, Guangxi, northern Zhejiang, southern Anhui, southeastern Hubei, Guizhou, etc., coal beds with economic value were formed in alluvial plain, transitional environment and back-barrier peat flat of shallow sea carbonate platform. In southwestern Sichuan, western Guizhou and eastern Yunnan, the Hsuanwei Formation on the margin of old upland, was formed in alluvial plain, having also economically important coal beds. The coal reserves of this period make up about 4.6% of the total. Although the coal reserve is not very rich, it is distributed in twelve provinces (autonomous regions), so that it may be one of three biggest platformtype coal-bearing areas of marine Permian.

The Late Triassic was a regressive stage and important coal-forming period in South China, such as Hunan, Hubei, northern Fujian and northern Guangdong. For example, the Anyuan Formation (Keuper and Rhaetian Stages), was formed in the transitional lagoon and bay environments. In several large continental basins, such as Yunnan, Sichuan, northern Shanxi, Qinghai and Tarim Basin of Xingjiang, there are all economically important coal basins formed in lake shore, swamps and flood plains.

There are also some coal basins with thin coal-beds distributed in Tibet, western Yunnan, western Sichuan, etc. The coal resources of this age only make up about 0.6%of the total amount in China.

The Early and Middle Jurassic was another important coal-forming period, in which there were many different-sized continent coal basins with different coal-bearing parameters, distributed in Northwest China, North China and the southern Northeast China. The coal basins, such as Tarim and Ordos, are all possessed of extra-large coal fields, with very rich coal resources, stable coal beds and simple geological structures. The other coalfields, such as Datong of Shanxi and Xishan of Beijing, are also important mining areas.

The typical coal-bearing formations of this epoch, the Datong Formation and Shuixigou Formation (Lias or lower Dogger), were formed in the lake delta and river environment, making up about 60% of the total coal resources of China.

The coal-bearing strata of the Late Jurassic-Early Cretaceous distributed in Northeast China and eastern Nei Mongol are represented by Jixi, Fuxin or Baiyinhua Group. They deposited in a series of down-warping faulted basins, which are separated with each other and synchronous. Many thick coal beds are often found in the strata. The coal resources of this age make up about 7% of the total amount of China.

There are also coal-bearing strata of Paleogene and Neogene, distributed in several provinces of East , Northeast and Southeast China. Palaeogene coal fields are primarily distributed in Northeast China, such as the famous Fushun coalfield, and so are Neogene coalfields in South Chins, such as the Xiaolongtan and Zhaotong coalfields in Yunnan Province. Thicknesses of coal beds usually are more than one hundred meters, and coal rank is lignite. The coal resources of this age make up about 0.8%of the total amount of China.

In addition, there are many peat fields of Quaternary in our country, particularly in Southwest China, Northeast China, eastern Nei Mongol and several coastal provinces. The amount of peat resources in China is about 700 hundred million tons.

2. Distribution of coal resources and coal ranks in China

By the end of 1986, our country's demonstrated coal reserve is 8 548 hundred million tons, and it is predicted that the coal reserve above the depth of 1 500m (1 000m in South China) is about 32 000 hundred million tons, making a total coal resources of 40 458 hundred million tons and occupying the front row among the world's coal-producing countries.

If the coal resources are arranged at provincial (regional) level, Xinjiang ranks the first, Nei Mongol the second, and Shanxi the third respectively; each with a total reserve of more than 5 000 hundred million tons. While the total coal reserve of another eight provinces (regions): Shaanxi, Ningxia, Gansu, Guizhou, Hebei, Henan, Shandong and Anhui amounts to more than 1 000 hundred million tons.

There are four large coal basins, namely Ordos, Turpan-Hami, Jungar and North China, with total reserve of more than 5 000 hundred million tons, characterized by large scale, thick and stable coal seams. Especially the Ordos coal basin is richly endowed with coal-bearing strata of Permo-Carboniferous, Late Triassic and Early-Middle Jurassic, with its coal resources of 18 000 hundred million tons, and its demonstrated reserve of about 3 000 hundred million tons.

In our country, there are widespread complete ranks of coal, ranging from lignite, bituminite to anthracite. The ratios of the different ranks of coal with total reserve in the whole country are: bituminite 83%, anthracite 9%, lignite 8%. If arranged according to metamorphic grades of coal, lignite accounts for 8% low-meta-bituminite coal 59%, medium-meta-vituminite coal 8%, high-meta-bituminite coal 6%, mixed coal 10%, anthracite 9%.

Of the total reserves, the coking coal is about 1/5, distributed mainly in Shanxi, Hebei, Mt. Helanshan, eastern Northeast China, Shandong, Henan, Anhui, northern Jiangsu, western Guizhou and eastern Yunnan.

There exist obvious regularities governing coal metamorphism. The coal of the Late Palaeozoic is primary mid-higher metamorphosed, while that of the Mesozoic is primary lower metamorphic bituminite. Only in eastern Nei Mongol, some lignite coalfields of the Late Jurassic were found. As for the coal of Cenozoic, only Fushun coal is bituminite, while the rest, such as the coal of Paleogene and Neogene in South and North China, are all lignites. Generally speaking, it reflects this trend: the earlier and the longer the coal-forming period is, the higher the rank of coal metamorphism reaches.

According to the coal rank, the lignites distributed primarily in eastern Nei Mongol, Northeast China and Yunnan, but in Guangxi, Guangdong and Shandong there are also some lignites. The anthracite is distributed mostly in central and southeastern Shanxi, central Henan and Guizhou, and southern Sichuan. The bituminous coal of lower metamorphism is distributed largely in Shaanxi, Ningxia, Gansu, Nei Mongol, Xinjiang, etc., whereas that of mid-higher metamorphism is primarily distributed in North China, East China and eastern Northeast China. The three-dimensional distribution of coal metamorphism has reflected that the deep burial metamorphism is the basis and has stacked with effects of telemagmatic metamorphism and dynamic metamorphism.

EXPLORATION AND EXPLOITATION OF COAL RESOURCES IN CHINA

The people of our country have found and used coal for as long as thousand years, and have accumulated rich experience in searching and mining coal. In the 1920s, Li Siguang, Wang Zhuquan, Xie Jiarong and other geologists made great contributions to the research on coal-bearing strata, geological survey and coalfield prediction. However, the regular exploration and exploitation of coal resources hadn't started until New China was founded. According to the requirements of industrialization, national economy and social development for energy resources, the Ministry of Coal Industry built up step by step a contingent for coal geological exploration, mine designing and construction. Under the guide of the long-term development plan of coal industry, we reconstructed old collieries first, and then undertook a major task to set up new coal industrial bases. In 1986, the output of raw coal reached 89 404 million tons, washed-coal 14 million tons. China has become one of the main coal producing countries in the world(Fig. 6).

Through geological prospecting for the construction of coal industry, the capability of coalfield exploration and scientific research has been developed and the technical competence improved. Most of coal geologic prospecting teams in the field have adopted comprehensively explorating methods, such as airborne and space remote-sensing, digital seismic, well logging, electromagnetic and gravitational methods, instead of single manner of geologic surveying and mapping in the past. Since the 1970s, thanks to the application of computers we have built up many kinds of data processing and testing centres for aerophotogrammetric mapping, remote-sensing image processing, seismic, electric prospecting and well logging. Also the preparations were started for setting up geologic data bases and making use of micro-computers to make geological reports. In recent years, by relying on the open policy, the Ministry of Coal Industry is continuously strengthening scientific and technological exchanges with friendly countries in the world. The Sino-American joint construction of Antaibao Opencast Mine in Pingshuo, the coal-field prospecting carried out by Chinese and Japanese technicians in Liuzhuang, Huainan, Anhui and Tangkuo, Jining, Shandong, were all done on the basis of equality and mutual benefit. We hope to develop the cooperative relations with all friendly countries and enterprises throughout the world.

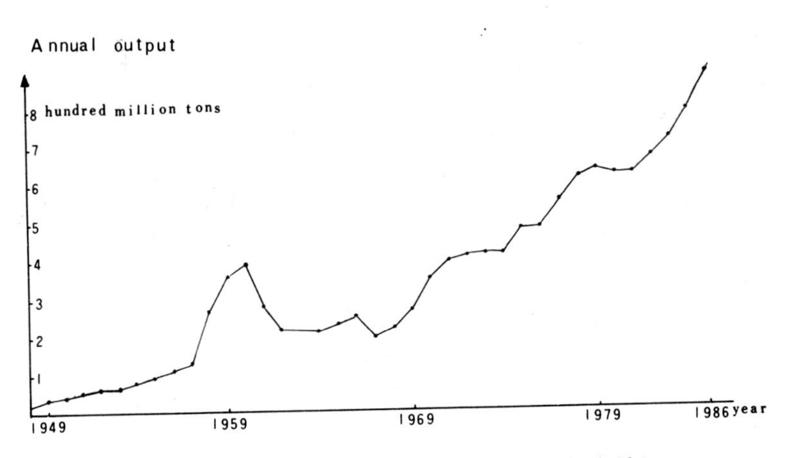

Fig. 6 Curve of annual output of raw coal of China

In the past 38 years, we have made remarkable achievements in coal exploration and exploitation. The three main aspects are as follows.

1. Many new coalfields have been found, and coal resources throughout the country have been basically ascertained

Coal mines in old China were small in scale, less in output, unavailable in coal resources and backward in mining methods. Right after the founding of New China, the former Ministry of Fuel Industry laid emphasis on reconstructing and expanding old collieries, conducting geological prospecting in coal mines and adjacent areas, so as to make the old mining areas rejuvenated and meet the needs of the economical construction of New China and the people's life for energy resources. In 1956, Prof. Wang Zhuquan and other geologists summed up experience in searching for coal in old mining areas, predicted new coal fields according to the distributive law of strata and structure conditions, launched geophysical prospecting and bore-hole drilling on large plain areas in East China. Only in a few years, they had found many covered coalfields in Shandong, Anhui, Hebei, Henan and Liaoning, etc. (Fig.1), which not only have thick coal seams, high quality and lauge quantity of coal, but also are near to Beijing, Tianjing, Liaoning, Nanjing, Shanghai and other coastal industrial areas, hence they have very high economic value.

After the large area geological mapping in exposed and semi-exposed coalfields from South, Northwest and North China to the west of the Taihangshan Mountains, quite a number of new coal fields have been found and ascertained. Since the 1970s, a large amount of lignite reserves have been recognized in Huolin He and Yiming, eastern Nei Mongol and Huang Xian of Shandong, and rich high-quality power coal reserves in Dongsheng, Jungar of southern Nei Mongol and Yushenfu, Bin Xian and Yuheng of Shaanxi in the well-known Ordos Basin by many of comprehensive new techniques and methods. With the construction of large opencast mines in Huolin He, Yiming, Jungar and the preexploitation in Dongsheng-Yushenfu coalfield, it is expected that a number of largest coal-producing bases will be set on the Northwest Plateau.

2. Some coal industrial bases have been built up and the multi-purpose utilization of coal has been developed

After the founding of New China, coal industry has ceen developing at a high speed. There are now about 100 coal industrial bases. Besides, in coal industrial system of our country there are small collieries run by counties, townships and individuals distributed in all provinces (regions).

As the number one energy resource in China, coal yields have increased by an average of 20 million tons a year during the past 38 years (Fig. 7). The increase of coal yields depends primarily on the development of coal-producing and constructing technology. The perfect technology of designing, mine constructing and surface building, especially that of sinking in special strata, have speeded up mine constructions. The yield of large collieries designed by us has come to 4 million tons a year as against original 900 thousand, 1.5 million and 2.4 million tons a year respectively. The opencast with a producing capacity of 10 million tons a year are under construction. Recently, the West Ventilating Shaft of Pansan Mine in Huainan, an overburden covered coalfield in East China, with the diameter of 9m and the depth of 508.2m, has been completed with an As-9/500 shaft drill of our own making, the diameter of the shaft being the biggest so far in the world.

On coal production, the underground mining is adopted as the main way in collieries of our country. In so doing, the output of hydraulic mining is 7-10 million tons a year, and the proportion of the opencast mining is only 4-5% of the total output. The sinking technique, the developing layout and the mechanization level of coal mine have been improved over the past 38 years. A number of mainstay collieries with modernized technology have been built up in recent years. In some mines where conditions of coal seams are complex and some senile ones, coal mining under three conditions, that is, mining under water, feeder railway lines, and buildings, is being successfully done. A large amount of coal are mined out, and the service life of mines increased.

Coal resources in China are very rich, but their distributions are not even. The demonstrated reserves and total coal resources are limited to the provinces south of the lower reaches of the Yangtze River. Energy resources are very important for economic prosperity in this coastal region of Southeast China, where the industry is flourishing and the population is dense. Our government has taken necessary measures to solve this problem step by step. We increase exploiting intensity for the producing collieries of Shandong, Hebei, Henan and Anhui, and speed up new mining construction to meet the needs of Shanghai and Jiangsu, Zhejiang and Fujian. In the meantime, we develop positively coal resources in eastern Yunnan and western Guizhou to meet the needs of Guangdong and Guangxi.

While increasing coal yields, we actively save on coal consumption and improve the utilization ratio of heat energy. We emphatically speed up constructing coal-washing plants for carefully processing coal in recent years. There are 140 coal-washing plants with various types, with a total processing capacity reaching 150 million tons a year. In the meanwhile, we develop the secondary transform of coal and new techniques of gasification, liquefaction, coking and gas-steam combined and circulated power generation. Coal is processed and utilized comprehensively on the spot and as nearby as possible according to different resource conditions. Some constructing integrated enterprises of coal, electricity, chemical industry, coking and gasification, are the effective way of making economic and rational use of coal resources.

3. Two times of coal resources prediction in the whole country have been finished for rational distribution of coal industry

We have finished two coalfield predictions for a long-term plan of national economy and the rational distribution of coal industry. The first one was in 1959. We based on known information of coalfield exploration, combined with regional geological data, ranging from one point to entire area, from the known one to the unknown one, inferred the deep from the shallow, and estimated coal resources by using the method of "from known mines to unknown mines" in Central, West and South China. In covered areas of East China, we used the method which combined geophysical prospecting with drilling proving and compared the area under exploration with known mine areas, and estimated coal resources. *Coal Geology of China, Coal Field Prognostic Map of China* and provincial atlases of that sort were published in 1960, which made coal geological theories of our country reach a new level.

The second coalfield prediction began in 1974. On the basis of geological works having been finished in main coalfields, the Ministry of Coal Industry worked out a policy "to sum up experience, explore regularities, ascertain resources, predict new coalfields". Hundreds of technologists from geological prospecting teams, mining bureaus, colleges and research institutes were organized to begin their work on mine areas prediction. Under the guidence of geomechanics theory, they studied the structures of mine areas, characteristics of sedimentary formations, distributive regularities of thick coal seam zones and prolific coal zones, drew different kinds of maps, summed up coal accumulation laws and conditions of coal preservation. The coalfield prediction in different parts of our country had been finished by 1979. After then, the Ministry of Coal Industry organized geologists to write new edition of *Coal Geology of China,* which had been published by Coal Industrial Publishing House in 1980,[7] had been awarded a prize of excellent scientific books of China, and has been translated into English to be published by Springer-Verlag. On the other hand, a number of coal geologists coming from different provinces (regions) and coal geological research institutes went in for coalfield prediction across the country. New edition of coalfield prediction, coalfield prognostic map, coalfield geological map, coal quality map of the People's Republic of China had been finished and published in series by the Ministry of Coal Industry in 1981, and won the top-grade prize of excellent achievements in science and technology. New edition *Coal Geology of China* and the achievements in coalfield prediction of China mentioned above marked a new stage of the academic level of coal geology in our country.

NEW ADVANCES OF STUDIES ON COAL GEOLOGY OF CHINA

1. Research in relation between coalfield existence and geological structure

Over the past 38 years, a large quantity of geological data have been collected in coalfield aerial reconnaissance and mining area exploration and development. Since the 1950s, some coal geologists have dedicated themselves to the study of aerial structural features of coalfield and medium- or small-scale geological structures which may affect exploration and development of mining areas. In the second nationwide calculation of coal resources during the 1970s, Li Siguang's theory of geomechanics that in a structural system and its compounding, the structural framwork has control on coal deposit area made coal geologists recognize some basic structural features of coalfield in China, and this recognization is more and more deepened and clear, because the application of the plate tectonic theory was expanded from the ocean to the continent when coal geologists were studying the lithofacies and palaeogeography of main coal-forming periods and the forming and evolution of coalfields in China(Fig. 7).

The complicated tectonic feature of China's mainland was formed on the base of the Early Proterozoic North China landmass which went through aggregation, amalgamation, accretion and collision with nearby landmasses and internal compression and tension as a result of Caledonian, Hercynian, Indo-Chinese, Yanshanian and Himalayan movements.

During the Palaeozoic Era, the North China landmass was a stable platform, bordered on the Pacific Ocean in the east and the ancient Tethys Sea in the west. During the Carboniferous and Permian periods, besides only thin coal beds deposited in the paralic molasse facies on the southern and western fringe of the landmass, important coastal plain and bay environmental Permo-Carboniferous coal measures were deposited on a broad Late Proterozoic and Early Palaeozoic carbonate platform under the background of Hercynian aerial subsidence. After the Late Permian, the North China landmass aggregated and amalgamated with the Siberian landmass in the north. The Mongolia-Tianshan folded belt was formed. Many terrestrial basins came into being within the North China landmass, while the Palaeozoic coal accumulation was wrapped up.

While the South China landmass was an epeiric sea during the Late Palaeozoic Era, there were some coal-bearing deposits on the border of upwarping areas during the Datang stage of Early Carboniferous. A wide regression occurred from the late Early Permian to the Late Permian, along with a major coal accumulation in the bay, coastal plain, delta and shallow sea carbonate platform environment and so the Permian coal deposits stretched over twelve southern provinces (autonomous regions). The South China and North China landmasses were aggregated at the end of the Hercynian period, and the Qinling folded belt was formed after the Middle Triassic. Thus, the three vast landmasses of North China, South China and Siberia had already aggregated together after the Indo-Chinese period to form the principal part of the Asian plate.

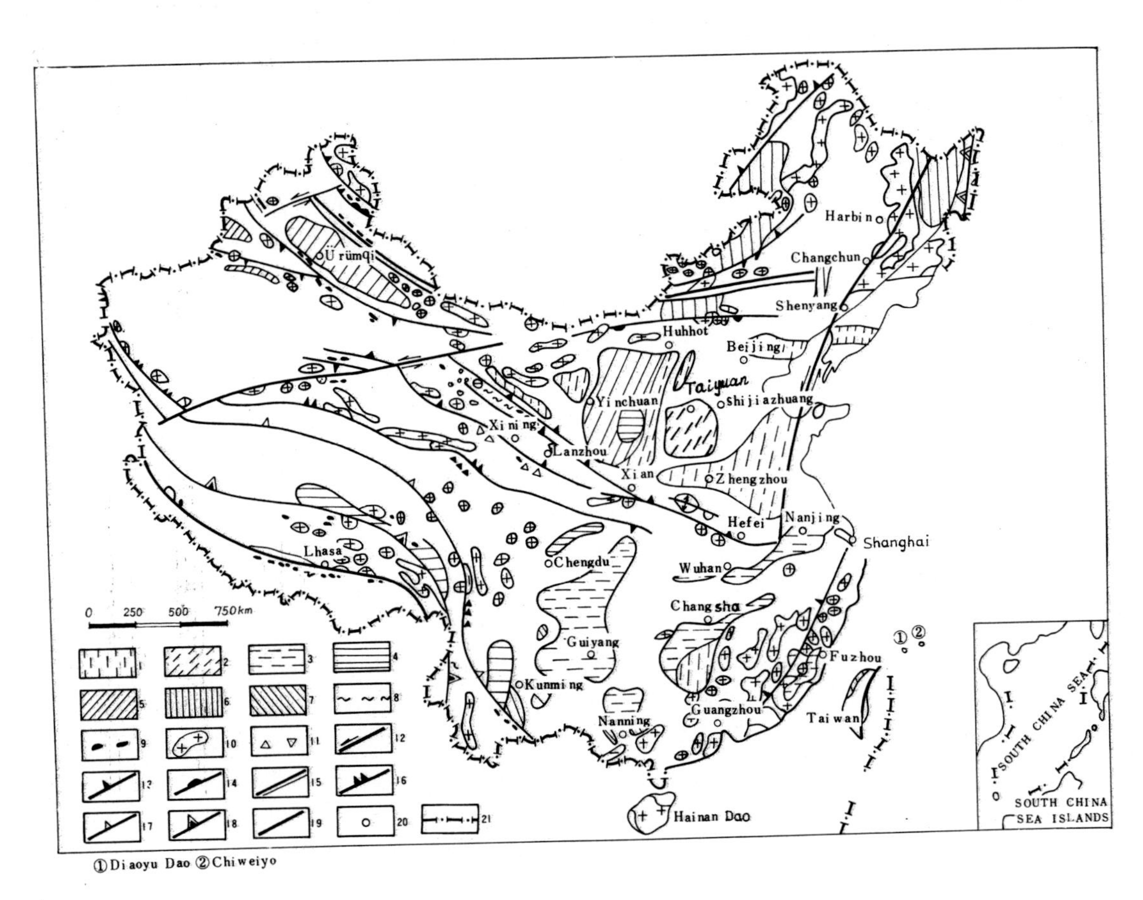

Fig. 7 Geologic tectonic map of coalfields in China

1. Coal accumulation areas (CAA) in Early Carboniferous; 2. CAA in North China in Permo-Carboniferous; 3. CAA in South China in Permian; 4. CAA is Late Triassic; 5. CAA is Early-Middle Jurassic; 6. CAA is Late Jurassic-Early Cretaceous; 7. CAA in Tertiary; 8. Glaucophane schist; 9. Ophiolite suite; 10. Synorogenic granite; 11. Melange accumulation; 12. Deep fault; 13. Caledonian subduction zone; 14. Early Hercynian subduction zone; 15. Middle-Late Hercynian suture belt; 16. Middle-Late Hercynian subduction zone; 17. Indo-Chinese subduction zone; 18. Yanshanian subduction zone; 19. Himalayan subduction zone; 20. City; 21. Border line.

During the Indo-Chinese and Yanshanian periods, the opposite movement between China landmass and Pacific-Philippines plates formed three upwarping and downwarping zones of northeast trend in East China. During the Cenozoic Era the Indian landmass in the southwest came into collision and then aggregated with the China landmass, meanwhile the Himalayan folded belt and the Qinghai-Xizang Plateau took shape. The huge faulted basins of Tarim, Qaidam, Junggar appeared in West China. As the stress field in

East China was tensile, thus the rift systems of Huabei, Songliao, Tanlu and Fenwei were formed. Along the 102-106^{0}E longitude, the north-south trend Helanshan, Liupanshan, Longmenshan and Daxueshan Mountains formed a natural border-line of two different tectonic types of East and West China.

Most Chinese coal geologists considered that the Hercynian and Indo-Chinese tectonic movements were decisive factors in the forming and distribution of the Late Palaeozoic coalfield, while the Yanshanian and Himalayan movements controlled not only the forming and distribution of Mesozoic and Cenozoic coalfields but also the reform of Late Palaeozoic coalfields. Generally speaking, the older the coal-forming age, the more complex the geological structure and the higher coal rank. The plate tectonics can satisfactorily explain the fact of coalfields in the three northeast-trend upwarping and downwarping belts, i.e. the structure is getting more complex and the metamorphism of coal is getting deeper from northwest to southeast.

The Ordos, Sichuan and Chuxiong coal basins are all large, broad depressed basins to form the westeast-northeast downwarping zone. The Permo-Carboniferous, Permian, Late Triassic, Early to Middle Jurassic coal-bearing strata formed nearly horizontal open folds. Fractures are rare, rather small, and devoid of igneous intrusion. The western North China upwarping zone occupied Shanxi, western Henan and northwestern Hebei. The Permo-Carboniferour and Early to Middle Jurassic coalfields are mainly medium- to small-scaled. Steep monoclines occupy the edge of the basins, while folds and faults are in the shallow part and broad or open folds in the interior part. Faults usually formed in the basin's fringe or shallow part in the form of graben and harst. The structures of coalfields on the South China upwarping zone in eastern Yunnan, Guizhou, eastern Sichuan, western Hunan and western Hubei are mainly box and comb folds, accompanied by larger or smaller strike faults. Hence these Late Permian and Late Triassic coalfields are rather complex in structure. Between Mt. Taihang and Tanlu fault lies the northeast-trend North China downwarping zone. The Permo-Carboniferous coalfields on this zone are mostly covered by younger strata. Fractures of different direction formed a series of lineament faulted blocks in different sizes. The structures of coalfields on this downwarping zone which stretches into South China, show a complicated picture of alternating depression and uplift belts in the directions of northeast and eastwest.

On the uplift belts, through a long period of erosion, the coal measures are only preserved within synclinoria. On the depression belts, the coal fields take the form of anticlines alternating with red basins. The northeast-trend anticlines are unsymmetrical, usually with northwest gentle limbs and southeast steep limbs. Low angle reverse faults were developed along the edge of some coalfields. The southeastern coastal Zhejiang-Jiangxi-Fujian-Guangdong upwarping zone is situated near the frontal zone of Asian plate. Since the post-Indo-Chinese crustal movements were very strong, not only are the structures of coalfields very complex, but also the igneous rocks destroyed many coal seams. And since the regional igneous rocks are extensively scattered, all the Permian and Late Triassic coals are paranthracite and superanthracite. In the coalfields in Datian-Longyan area, Fujian, very complex structures such as Z-type folds and nappes and gliding nappes are well developed, known as thus the most complex coal fields in the world. The R^0_{max} value of coal from Shangjing mine of Datian, is 8.02%, while that of the Triassic coal from Jianou, northern Fujian is 7.42%, both being superanthracite.

There are two different structure types of the Late Jurassic coalfields in eastern Nei Mongol and Northeast China. The structures of Hegang, Shuangyashan, Jixi and Qitaihe coalfields on the northeastern Heilongjiang uplift are mainly faulted blocks, often with coal seams there being destroyed by igneous rocks. Here the coal mainly belongs to the medium-metamorphosed bituminous coal, while the coal from both sides of the Da Hinggan Mountains are lignite. As often as not, hundreds of medium and small coal basins in our country are graben or graben-like faulted depression basins.

As already mentioned above, the major bases of coal industry are located in the eastern part of China. In this area, exhaustive studies of regional geology were made and coalfields here were also well prospected and developed. During the last thirty-odd years, a lot of new coalfields were found, and a search for more new coalfields is getting under way. In recent years, Chinese coal geologists applied the plate tectonic theory to the prospecting of the coal in and near mining areas, which resulted in remarkable success in western Henan (Fig. 8), Huainan of Anhui (Fig. 9), Longyan of Fujian, Tangshan of Hebei, etc. Because new coal resources were found under the thrusts and gliding nappes, the old mine areas were rejuvenated in East China.

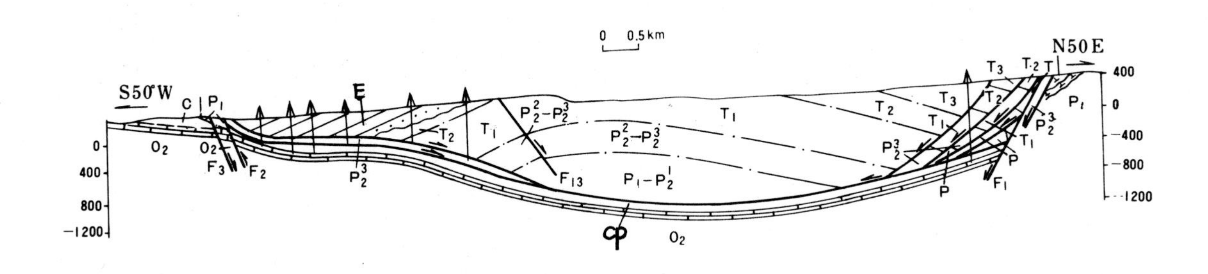

Fig. 8 Structural cross-section of Ludian, Henan

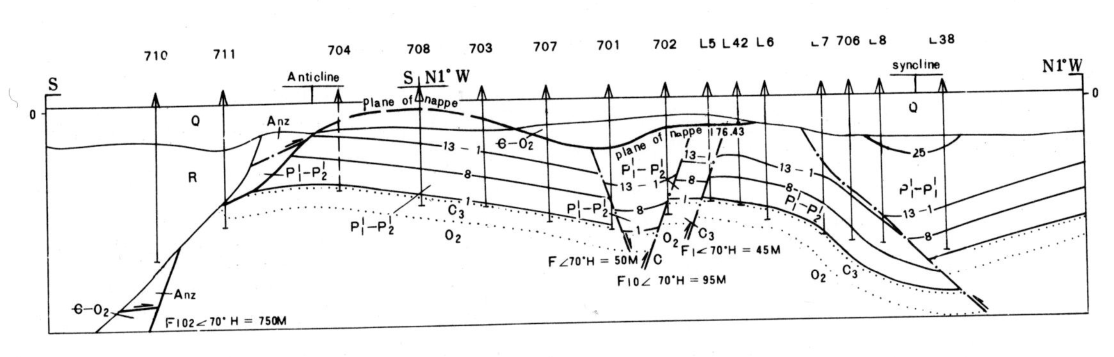

Fig. 9 Structural cross-section of newly found coal-bearing area of Huainan, Anhui

2. Research on coal-bearing strata and palaeontology

The stratigraphical division and correlation of coal-bearing strata are the basis for the research of sedimentary formation and coal-forming regularities. Originally, most coal-bearing strata in China were named after local names. During the past thirty more years, a large amount of systematic work have been done on division and correlation of coal-bearing strata by Chinese coal geologists. Until the 1960s, the chronostratigraphic classification of Palaeozoic coal-bearing strata was basically unified, and a unified correlation scheme of coal-bearing strata of different ages was drawn out. Owing to the research of coal-bearing strata, a large amount of fossil especially microfossil data were accumulated, for example, the discoveries of *Sphaerulina-Nankinella* and *Staffella* Zones above the *Pseudoschwagerina* Zone of the former top of Taiyuan Formation in northwestern Henan and eastern Nei Mongol, and sponge spicules and *Lingula* sp. in many places in the originally believed "terrestrial" Shihhotse Formation of North China. Another example is the study of *Gigantopteris* flora of Permian coal measures in South China, from which a conclusion was made, i. e. in South China, from the late Early Permian to late Late Permian there was a continuous process of coal-forming. And in western Guizhou, a mixed flora of *Gigantopteris* and *Glossopteris* was found. Because of a relatively correct determination and correlation of coal horizon of various ages, a foundation was laid for the study of coal-forming regularities, coal resource estimation and new coalfield prediction.

3. Studies on depositional environment and coal-forming regularities

The subjects of main coal-forming periods and coal-forming regularities are always the study focuses of coal geologists. In recent years, the theories of depositional environment and sedimentary facies were used in searching for coal-forming models in different depositional environments. Among the outstanding examples are the Late Jurassic coalfields of Fuxin of Liaoning, Huolin He of Nei Mongol; Permo-Carboniferous coalfields of Taiyuan, Xishan, Yangquan, Shouyan of Shanxi Province and Yu Xian of Henan; Permian coalfield of Heshan of Guangxi and southern Sichuan; and Jurassic coalfield along Mt. Qilian. Geologists at large have made a vast amount of field depositional environmental studies, worked out various coal-forming models of different depositional environments, such as delta, back barrier peat flat and faulted basin, achieved a series of positive results during coal exploration and exploitation and coal prediction. Moreover, in North China, the setting up of tidal flat model of Taiyuan Formation of western Henan, Huainan and Huaibei coalfields and transitional environment of Shihhotse Formation has provided important basis for the study of coal-forming regularity.

1) Back barrier peat flat coal-forming model

In central Guangxi, southeastern Guizhou, eastern Sichuan, southern Hubei and northern Jiangxi, the main coal-bearing strata are carbonate-dominated Hoshan or Wuchiaping Formation of Late Permian, in which limestones make up more than 85% of the total amount of rocks, and which contains a great quantity of fossils anf fossil fragments, with many beds or members consisting of biolithites. Coal seams were formed in a tidal peat flat environment.

On carbonate platform in epeiric sea, some bioherm deposits seem to have nothing to do with coal seams, but they formed barriers, with a large number of plants living on the back-reef flat, which turned into peat flat and later became minable coal seams (Fig. 10).

In Heshan coal mine of Guangxi, coal seams are less in quantity, but stable in coal horizons, variable in thickness, high in ash and sulfur content and complex in coal seam structure, known as a new kind of coal-forming environment in the world, and with theoretical significance.

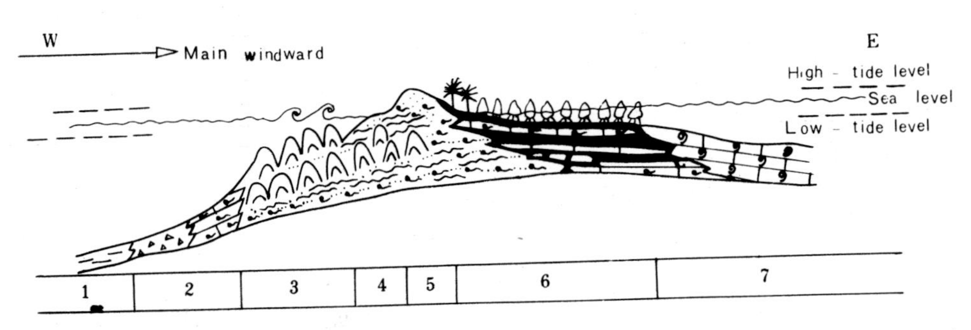

Fig. 10 Sedimentary models of organic reef and peat-flat in Hoshan Formation of Matan region

1. Open marine facies; 2. Fore-reef facies; 3. Reef-core facies; 4. Reef-flat facies; 5. Lime-sand island facies; 6. Back-reef facies (consist of back-reef peat flat facies); 7. Restricted subtidal facies.

After having studied depositional environment of coalfield in Matan area, central Guangxi according to the difference between settling amplitudes, Zhang Pengfei and Liu Huanjie divided the back-barrier peat flat coal-forming model into two types: the first progradational sequence: localized subtidal facies—intertidal facies—mixed flat facies—peat flat facies; retrogradational sequence: peat flat facies—intertidal facies—localized subtidal facies; the second progradational sequence: localized subtidal facies—intertidal facies—tidal channel facies—peat flat facies; retrogradational sequence: peat flat facies—intertidal facies—localized subtidal facies—reef facies.

2) Coal-forming model of the Mesozoic and Cenozoic faulted basins

The Mesozoic and Cenozoic faulted basins of our country are composed mainly of the Late Jurassic and Early Cretaceous coal basins of Northeast China and Nei Mongol and the Tertiary coal basins of eastern coastal provinces and Yunnan. The former are a series of faulted lake basins formed during the Yanshanian movement beginning with vast volcanic eruptions. There are about 100 basins, with the general depositional sequence given below:

(1) Basal pluvial-alluvial coarse clastic sediments;

(2) Lower coal member (basin margin) or lacustrine mudstone member (basin center);

(3) Lacustrine mudstone member;

(4) Main coal member;

(5) Top pluvial-alluvial sediments.

In some of the basins, the repeated members of (2) and (3) made the lower coal member extremely thick, such as the Damoguai Formation of the Yimin Basin in eastern Nei Mongol. Generally, there are thick coal seams in the main coal members suitable for openpit mining because they are shallow and very thick (Fig. 11).

Depositional environment and tectonic movement affect the thickness and geometry of coal seams. Coal seams split and pinched-out usually near the fans, thinning down at the border of the deltas. Usually coal was formed at the shallow part of the basin, and pinched out toward the deep lake in the basin center. Most appropriate coal-forming environment is the lake shore depression between them and usually the basement fault remobilization during the Himalayan movement causes the thickness of the coal changing abruptly right upon the fault.

The latter (i. e. the Tertiary coal basins) which carried on the old fault mobilization during the Himalayan movement, accumulated coal beds in the intermontane basin of the negative graben of faulted zone with the evolution of the interior basin (Fig. 12). The Tertiary coal-forming period can be divided into two stages. In Eocene, there are Xialiaohe, Fushun, Shulan, Meihe coalfields in Northeast China, and Huang Xian coalfield in Shandong Province, in Maoming, Changchang of Guangdong; in Taiwan, there are also some small coalfields belonging to this stage. The coal-forming process during Neogene period mainly happened in Yunnan (Fig. 13) and Guangxi, consisting of hundreds of small-scale coal basins. The general depositional sequence is:

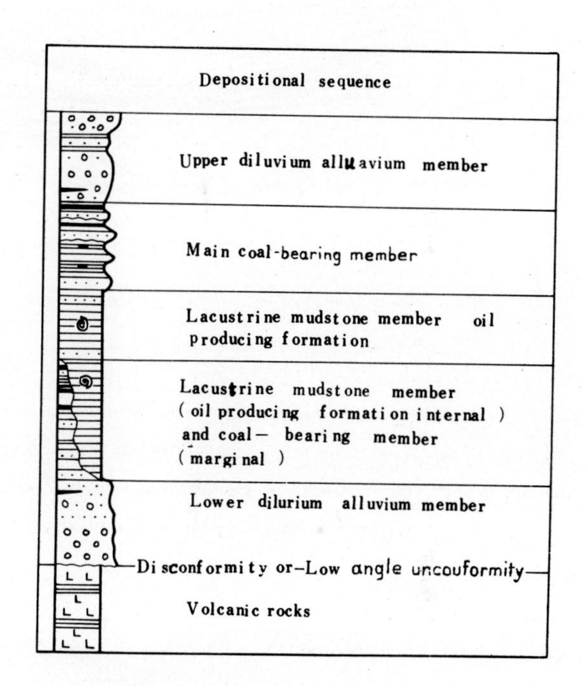

Fig. 11 Depositional sequence model of Late Jurassic-Early Cretaceous faulted basins of Northeast China and eastern Nei Mongol (after Li Sitian)

(1) Basal coarse clastic rocks;

(2) Thick coal or coal intercalated with sandstones and shales;

(3) Oil shales, mudstones (diatomaceous earth), being characterized by almost having no relation with volcanic eruption.

But in eastern Zhejiang and Fujian, Shanxi, northern Hebei and Nei Mongol, the coal-forming process usually happened between the intermittent eruptions and intercalated with basaltic flows, but the coal is thin, as seen in Fanshi of Shanxi and Sheng Xian of Zhejiang.

4. Studies on coal metamorphism in China

The coal metamorphism is a theoretical problem to which coal geologists of our country have been paying close attention. It has been proved that the hypometamorphism of coal exists generally in coalfields all over the country, through the regional geological work, exploration and exploitation, and the prediction of coal quality in the past thirty-odd years. The different subsidence ranges during and after coal-bearing strata were deposited, exerting very considerable influence on the distribution of coal metamorphic zone (Fig. 14).

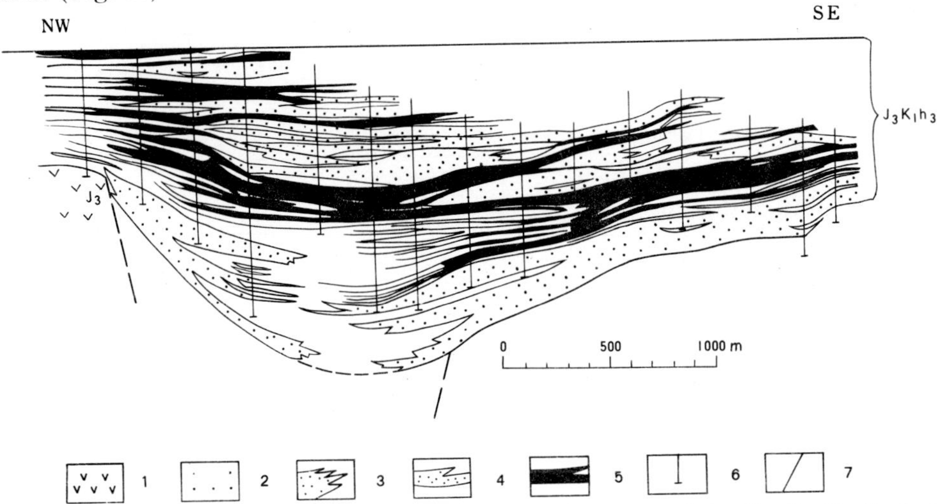

Fig. 12 Prospecting line depositional cross section of northern part of the Huolin He Basin (Vertical scale is 2.5 times exaggerated)

1. Medium acidity volcanic rocks of Hingganling Group; 2-5. Huolinhe Group: 2. Basal arenaceous-conglomeratic member; 3. Fanglomerate; 4. Sand-body; 5. Coal seam; 6. Drillhole; 7. Fault.

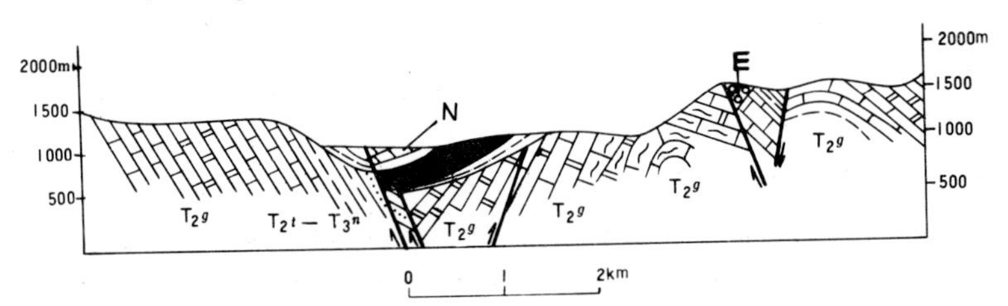

Fig. 13 Geological cross-section of the Xiaolongtan Basin, Yunnan (after Hu Youheng)

However, most China's metamorphic coal widely appears though the coal-bearing and overlying strata are not thick. Coal geologists had two lively discussions on main factors of China's coal metamorphism late in the 1950s and 1970s. In 1956, Prof. Wang Zhuquan proposed "magmatic metamorphism", and published a paper *Zonal distribution of coal ranks and its geological factors in North China,* in which, he pointed out "Though the regional metamorphism and dynamic metamorphism exerted some influence on coal ranks, the metamorphism of igneous rocks played a leading role in North China." "The granitic intrusive masses are widespread in Fujian, Zhejiang, Guangdong and eastern Jiangxi, so coals from these regions are chiefly anthracites whether in Permian or in

Triassic". The views mentioned above have been supported by many coal geologists. Since 1970s, the artificial coalification tests by some geologists, the studies on the properties of maceral and on the coal-bearing strata alteration by igneous rocks, and the work on coal quality zoning, have further proved that the main factor of coal metamorphism is magmatic metamorphism. It is a breakthrough to the traditional concept of contact metamorphism, and enriches the coal metamorphic theory of our country (Fig. 15).

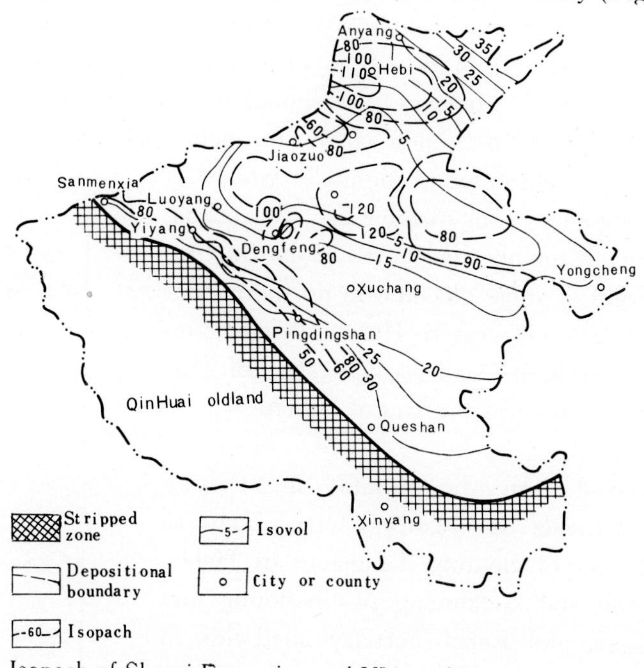

Fig. 14 Isopach of Shansi Formation and V^r isovol No.=-1 coal, Henan.

In the meantime, some geologists, according to the theory of geomechanics, think that coal-bearing strata are carrying different kinds of crustal stress. When the lateral compression and shearing reach a certain degree, the heat produced by internal friction among rock grains is conducted and accumulated in a certain way and as a result, becomes the main factor of coal metamorphism, a new viewpoint called "tectonic stress metamorphism" which is the main factor of coal metamorphism in China.

5. Studies on the special coals of China

The special coals, such as "stone-like coal", "sapropelic coal" and "bark coal", have a certain significance to China's coal resources.

1) Stone-like coal

Stone-like coal is widespread in northern Sichuan, southern Shaanxi, southwestern Henan of eastern Qinglin Mountains and Zhejiang, Jiangxi, Hubei, Hunan, Guizhou, etc., surrounding the Jiangnan upwarping, and was formed in Cambrian-Silurian period of the Early Palaeozoic Era. The ash content of common stone-like coal is 40-90%, with caloricity under 4 000 cal/g; and that of high quality stone-like coal is 20-40%, with caloricity 4 000-7 000 cal/g. The total resources of stone-like coal reach over 1 000 hundred million tons and the common coal is the most abundant; both can be used not only as domestic fuel, but also as raw materials for some small-scale industry, such as power generation, building materials, gasification and chemical fertilizer as well as vanadium, molybdenum, nikel, uranium, copper, etc., extracted from stone-like coal in coal-lacking area of South China.

Stone-like coal was formed in the shallow sea environment around the oldland during the Early Palaeozoic Era. It is a kind of sapropelic coal which came from some lower organisms such as alga in the stagnant water and reduction environment, with a nitrogen-hydrogen ratio (H/N) of 6-7. It is proved from extracted bitumen and cluster analysis that their original substances are similar to those oil-forming substances with a reflectance (Ro) of 13.4-17.5%. So it is called sapropelic anthracite.

In addition the "bituminous lignite", which is associated with stone-like coal in the Early Palaeozoic Era is South China, is mainly distributed in the fracture zones on the margin of some old land in northern Zhejiang, southern Anhui and western Hunan, locally minable in some small collieries such as Kangshan of Zhejiang, Taiping of Anhui and Dongkou of Hunan.

2) Sapropelic coal

Sapropelic coal can be found in Permo-Carboniferous coalfields in North China, such as Hunyuan and Huoxi of Shanxi, Tangshan of Hebei, Feicheng, Yanzhou and Zaozhuang of Shandong and Xuzhou of Jiangsu, and Early Tertiary coalfields in Fushun of Liaoning, Huang Xian of Shandong. They usually occur as the thin distinct seams between main minable coal seams or as individual seams on the roofs or floors of coal seams (generally 30-40cm) associated with humic coal and oil-shale. Sapropelic coal has dull bitumen luster, conchoidal fracture, inflammable, low specific gravity (1.21-1.27), etc. Tests proved that the original substances are chiefly algae (about 23-80%). This kind of coal belongs to sapropelic coal or sapropelic-humic coal formed in deep lakes, associated commonly and closely with oil shale. Laterally, there are an associating succession of lacustrine sapropelic coal, shallow lake oil shale and carbonaceous shale, in addition to humic coal formed in lake shore peat swamps. Vertically, from bottom to top, the associating succession is composed of sapropelic coal, oil-shale, carbonaceous shale, seat earth and humic coal.

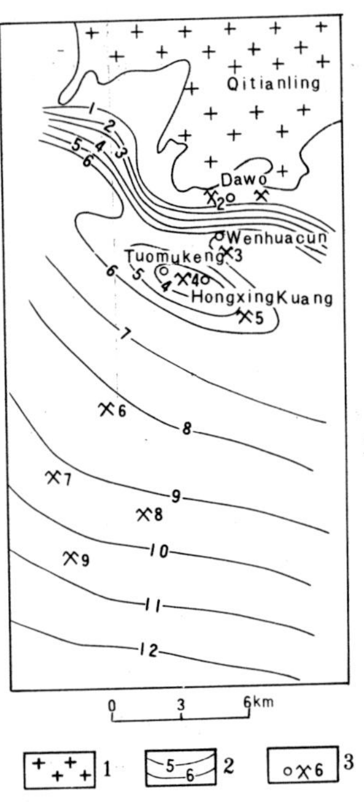

Fig. 15 A sketch showing relation between Qitianling granite and isovol of No. 12 coal seam

1. Indo-Chinese granite; 2. Isovol of No. 12 coal; 3. Sampling site and its number.

Besides, "jet" can be found in the coal seams in Fushun coalfield of Liaoning, known as a kind of mixture of sapropelic coal and humic coal, which can be used as material of artworks.

3) Bark coal

Bark coal, a typical coal in the Late Permian in South China, has been well-known in the world since Prof. Xie Jiarong studied and gave a name "Leping coal" in 1933. Now it has been ascertained that it is distributed on the southern margin of Jiangnan old upwarping from Yixing of Jiangsu, Changxing of Zhejiang, Guangde of Anhui, Leping, Gao'an and Wanzai of Jiangxi, to Ningxiang and Liaoyuan of Hunan. Besides, it is also dispersed in Shuicheng and Pan Xian of Guizhou, and Xuanwei and Fuyuan of Yunnan.

Bark coal has greasy luster, bitumen smell, high toughness and is inflammable. It is defined as a varied maceral type with rich bark by the studies of coal petrology. Among maceral groups, the content of suberinite reaches to 60-90%, valatile matter 5% and hydrogen 4.7-6.86%. According to the information from Changguang coalfield, the spoils in coal seams contain a number of marine animal fossils, such as Foraminifera, Echinoidea, Brachiopoda and pecten. The strontium-barium ratio (Sr/Ba) reaches 2.67%. It was formed under the environment of the biodetritus beach ridge-tidal flat or peat flat in the bay.

CONCLUSIONS

China is one major coal-producing countries in the world and also a big country using coal as primary energy resources which resulted from China's developing energy resources industry and the objective reality. Because of these, the exploitation and exploration of coal resources are very important in the national economy and social development. Coal yields of China will come to 14 hundred million tons and the technology of state-owned coal mines will reach or approach the same level as the advanced countries by the year 2 000.

Since the establishment of New China, a top priority has been given first to exploration and exploitation of the eastern coalfields where industries and population are densely distributed. Great efforts have been made over the past thirty years and the coal resources have been ascertained and vast amounts of coal reserves are under mining and utilizing in Northeast China, North China and south of the lower reaches of Yangtze River. At present, the mining depth of large coal mines has come to 600m. With the increase of deep mining conditions and fully mechanized coal facies, the accuracy of exploration result must be raised. The exploration and exploitation of deep and thick overburden covered coalfields become more difficult, which demands a higher level of exploring and mining technology. Nowadays, the focal point of exploration and exploitation is gradually shifted to new energy resource bases primarily in Shanxi. Large scale constructions have been set off in the coking coalfields and those coalfields with high quality, low ash and sulphur power in Shanxi, western Henan, northern Shaanxi and southern Nei Mongol. A number of new coal industrial bases are taking shape.

Coal resources in West China are very rich. For example, the total amount of coal resources of Xinjiang is more than 10 000 hundred million tons, but the demonstrated reserves make up only 4/1000 of total demonstrated reserves in China. The degree of exploration in this region is much lower than that in the eastern part of China and coal mining is limited along the railways and around the cities. In the famous districts of Jinshajiang, Nujiang and Lanchangjiang Rivers (northeastern Xizang, western Sichuan and Yunnan as well as southern Qinghai) the Late Tertiary marine and terrestrial alternating facies strata exist widely, with only little geologic work being done.

As stated above, the working field of our country's coal geology is very extensive. Coal geologists should continue to sum up their experience actively and study the new theories, methods and techniques both from home and abroad. With the carrying out of coal geologic work and strengthening of coal geologic studies, better geologic achievements will bring about a better service to the coal industry. The more glorious future of coal industry and coal geology will be about to come!

This report is compiled by Geological Bureau and Institute of Geology and Exploration, Ministry of Coal Industry. Members of the compiling group consist of Wang Zhongtang, Pan Suixian, Mao Bangzhuo, Li Wenheng and Guan Shiqiao.

REFERENCES

1. Coal Geological Institute of Sichuan, Stratigraphic Subdivision and Coal-bearing Characteristics of the Late Permian Epoch in South Sichuan, Chongqing Press (1986).
2. Geological Explorative Institute of Chinese Academy of Coal Science and Coal Explorative Company of Shanxi, Sedimentary Evironments of Taiyuan Xishan Coal Basin, Shanxi Province, Coal Industry Press (1987).
3. Han Degin and Ren Deyi, Acta Sediment Sinica, 4 (1983).
4. —— and Yang Qi, Coal Geology of China, Coal Industry Press (1980).
5. Han Zhiwen and Yang Qi, In: Proceedings of the Symposium on Permo-Carboniferous Coal-bearing Strata and Geology, Science Press (1987).
6. Huang Jiqing, Acta Geol Sinica, 40 (1) (1960).
7. Li Chunyu, Ibid, 57 (1) (1983).
8. Li Sitian et al, Ibid, 56 (3) (1982).
9. ——, Earth Science, 18 (3) (1982).
10. Li Tairen, Jour Coal Soc China, (2) (1983).
11. Li Tingdong, Jour Chin Acad Geol Sci, (4) (1982).
12. Li Wancheng, Coal Geol Expl, (2) (1979).
13. —— and Su Jinping, Ibid, (2) (1986).
14. Miao Fen, Sci Bull, (3) (1980).
15. ——, In: Geomechanic Collection, Geological Publishing House (1987).
16. Ministry of Coal Industry, Coalfield Prediction of People's Republic of China, Coal Industry Press (1981).
17. Ministry of Coal Industry, In: Coal Industrial Yearbook in China, Coal Industry Press (1982).
18. Pan Suixian and Lin Ji, In: Scientific Papers Collection of International Geological Exchange (3), Geological Publishng House (1980).
19. Qian Lijun and Bei Qingzhao, The Stratigraphical Division and Correlation of Mesozoic Era Coal-bearing Strata and Its Palaeogeography in Southern China, Coal Industry Press (1987).
20. Wang Hongzhen, In: Atlas of the Palaeogeography of China, Cartographic Press (1985).
21. Wang Rennong, Coal Geol Expl, (5) (1982).

22. Wang Zhongtang, Acta Geol Sinica, 45 (1) (1965).
23. Wang Zhuqian, The Forming Conditions and Distributive Regularities of the Late Permian Coalfields in South China, Coal Industry Press (1980).
24. Xiu Fuxiang, Coal Geol Expl, (1) (1984).
25. Yang Qi, Earth Science, 18 (3) (1982).
26. Yang Qi and Ren Deyi, Coal Geol Expl, (1) (1981).
27. Yao Zhaoqi, Acta Palaeont Sinica, 17 (1) (1978).
28. Yuan Yaoting, Coal Geol Expl, (3) (1978).
29. Zhang Pengfei and Liu Huanjie, Acta Sediment Sinica, 1 (3) (1983).
30. Zhang Renbao and Sang Shaohua, In: Proceedings of the Symposium on Permo-Carboniferous Coal-bearing Strata and Geology, Science Press (1987).
31. Zhang Wenyou, Sci Sinica, (2) (1978).
32. ——, An Introduction to Fault-block Tectonics, Oil Industry Press (1984).

XIe Congrès International de Stratigraphie et de Géologie du Carbonifère Beijing 1987,
Compte Rendu 1 (1991): 142-156

CARBONIFEROUS SEDIMENTARY TYPES IN CHINA WITH REFERENCE TO PETROLEUM POTENTIAL

Tian Zaiyi and Tang Wensong
(Research Institute of Petroleum Exploration and Development, Ministry of Petroleum Industry, China)

Carboniferous strata of China are developed not only as a coaly sequence but also an important one for hydrocarbon formation in geological history. In the East Sichuan Basin, several natural gas producing horizons in Middle Carboniferous have been verified in Xianggousi, Leiyinpu and Zhangjiachang structures. In Wenliu, Suqiao of the Bohai Basin and Zizhou, Suide areas of the Ordos Basin, pools of coal gas derivcd from Upper Carboniferous have been found. Meanwhile, in other vast areas such as East Yunnan, South Guizhou, Hunan, Guangxi, the Gansu Corridor and the Tarim Basin, there are a lot of oil, gas and asphalt shows. It may be inferred from the productive gas and different types of oil or gas shows as presented above that the Carboniferous strata of China are substantial source rocks for oil gas generation and naturally, have aroused great attention to petroleum geologists.

Carboniferous strata of China are of complicated types of sedimentary facies varying from marine, transitional to continental facies. Apart from North China where there is commonly a lack of its lower part, Carboniferous crops out completely in other extensive regions of China. Based on their sedimentary assemblage, they may be divided into: (1) the stable epicontinental sea type as in North China, Tarim and Yangtze Platforms; (2) the active pelagic type as in Tianshan-Mongolia-Hinggan Ling Trough, Kunlun-Qinling Trough and West Yunnan of Tethys Sea; and (3) the transitional type in areas as those between the first and the second types.

DISTRIBUTION

1. North China stable sedimentary region

Uplifted since Ordovician, the North China platform hadn't had any sediments deposited for a long period of time. In the later period of Early Carboniferous, transitional continental clastics of huge thickness filled only at the foot of Dabieshan Mountains.[1] It was not until Middle Carboniferous when the crust depressed, that the sea water from Taizihe River,[1] northeast of the platform, transgressed gradually over North China and the Penchi Formation was formed. In the lower part of the Penchi Formation are the so-called "Shansi-Type " hematite and bauxite because of the accumulation of elements of iron and aluminium on the erosional surface, and upward come sandy shales, black shales, limestones and coal beds which represent a coastal and swamp environment. In Late Carboniferous, more far-reaching transgression led to the overlap in Ordos, South Henan and North Anhui. The Taiyuan Formation which was deposited in alternate marine and continental background is characteristic of three sedimentary cycles, each of which consists of coarse-grained massive quartz sandstones in the lower part and shales, coals interbedded with limestones in the upper part. The formation is 10-250m thick, regionally stable and was overlain by the Lower Permian with an unconformity (Fig.1). In the west margin connecting the platform with Gansu Corridor, the situation is a little different. The area experienced a long period of erosion in Devonian. In Early Carboniferous, the area became a coastal sea or lagoon in which sandstones, shales and limestones were interbedded with gypsum and coal layers, 100-500m in thickness. The middle of the formation includes notably nonmarine clastics and marine mud-limestones, with a thickness of 200-300m. The upper part of the formation gives way alternately to marine/nonmarine clastic rocks, limestones and siliceous rocks, with a total thickness of 40-280m.

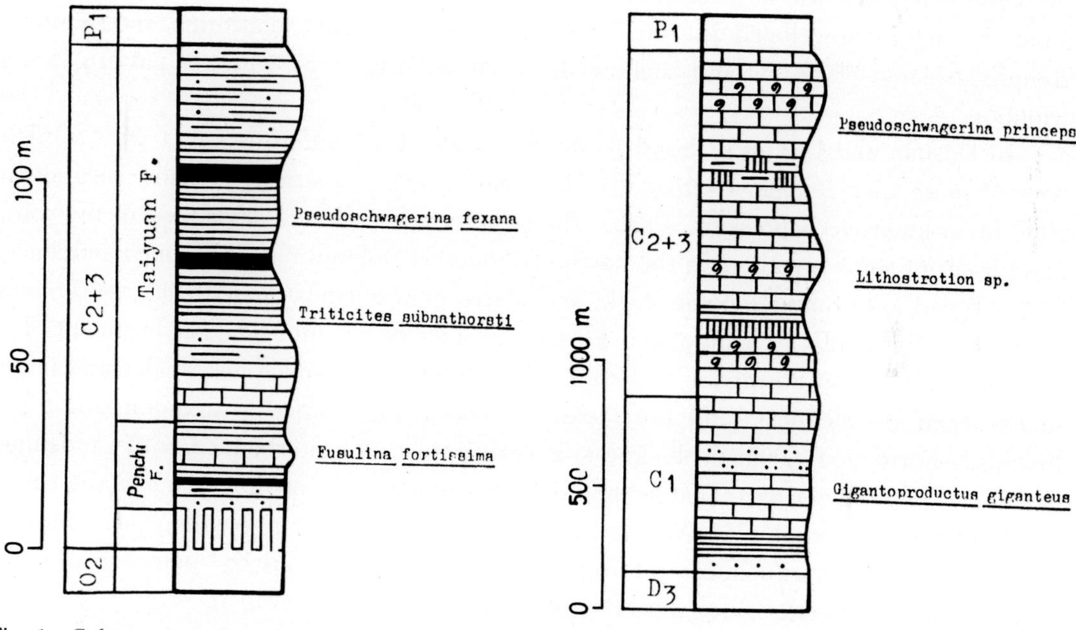

Fig. 1 Columnar section of Carboniferous strata in Taiyuan, Shanxi

Fig. 2 Columnar section of Carboniferous strata in the Tarim Basin

2. Tarim stable sedimentary region

With the observation of exposures in Kalpin, Tiekelike and Kuluktag in the periphery and the correlation by means of seismic and drilling data within the basin, it seems reasonable that Carboniferous strata are distributed widely in the whole Tarim Basin. In Kalpin, the Lower Carboniferous strata, 400-1 000m thick, contain sandstones, shales in the lower group and limestones interbedded with fine-sandstons and thin layers of localized gypsum in the upper group. In Tiekelike, the sedimentation was much the same and massive limestones in the upper part might result from the sedimentation of subtidal and intertidal environments. In the west of the platform, the Middle-Upper Carboniferous section, 1 400m thick, consists of shallow marine carbonates with minor amounts of reefs, gypsum-bearing mudstones, gypsum layers and siderite, coal-bearing mudstones. In Kuluktag, Middle Carboniferous clastics and carbonates deposited in shallow sea, 1 400-2 000m in thickness, but the area was folded up just after then. In short, the Carboniferous strata in Tarim Platform are steady in both facies and thickness and may be classified as sediments in an open sea platform(Fig.2).

3. South China stable sedimentary region

The area investigated includes the Yangtze Platform and the post-Caledonian stable region. The Liujiang Movement in Late Devonian had an intensive influence on the areas in South China and the sea water waned back from most of the region except for two remaining seas in Central Hunan and South Guizhou, so that the Early Carboniferous was short in sedimentation. There were frequent fluctuations and migrations in marine and continental geography in Carboniferous time, but generally speaking, along the margin of the continent there was a swamp which is favourable for coal formation and seaward from the margin developed the facies of continental margins, platforms, interplatform troughs, isolated platforms and shallow/deep basins, usually in a proper pattern in distribution.

In Dushan and Duyun of South Guizhou, Lower Carboniferous strata with a thickness of more than 2 000m overlay the Devonian strata uninterruptedly and were deposited in an intensive depression, so that the strata are completely developed in the south and absent of some members in the north. The muddy dolomitites, limestones and shales of the Kolaoho Formation, 150m thick, and marls, quartz sandstones, shales and dolomitites, 160m thick, of the Tangpakou Formation indicate an intertidal environment. The coal-bearing Chiussu Formation is dominantly composed of sandstones, shales and limestones, 300m in thickness, and the facies are marine alternated with nonmarine. The Shangssu Formation, 600m in thickness, is carbonate rocks interbedded with sandstones and shales in its lower part. Due to the relatively minor rising of the crust and the subsequent gentle topography in Middle-Upper Carboniferous, the Weining and Maping Formations with a thickness of only 400m and characterized petrologically by homogeneous carbonate facies were deposited in shallow water (Fig.3).

4. Xizang (Tibet) stable sedimentary region[2]

The region is extensive and the Carboniferous strata are widespread in distribution and complete in section. 2 000-3 000m in thickness, representing largely a stable sedimentation. In Himalayas, clastic rocks formed in shallow water and from bottom to top are divided into Yali, Naxing and Jilong Formations. They consisted of black shales with sandstones, marls and conglomerates. In North Xizang (Tibet), the strata of Lower Carboniferous have several members of chert-bearing limestones, marls, bioclastic limestones, shales, sandstones and coal layers, totally 2 270m in thickness. The Middle Carboniferous is composed of limestones, 60m thick. And the Upper Carboniferous consisted of limestones, dolomitites with shales, 360m in thickness (Fig. 4).

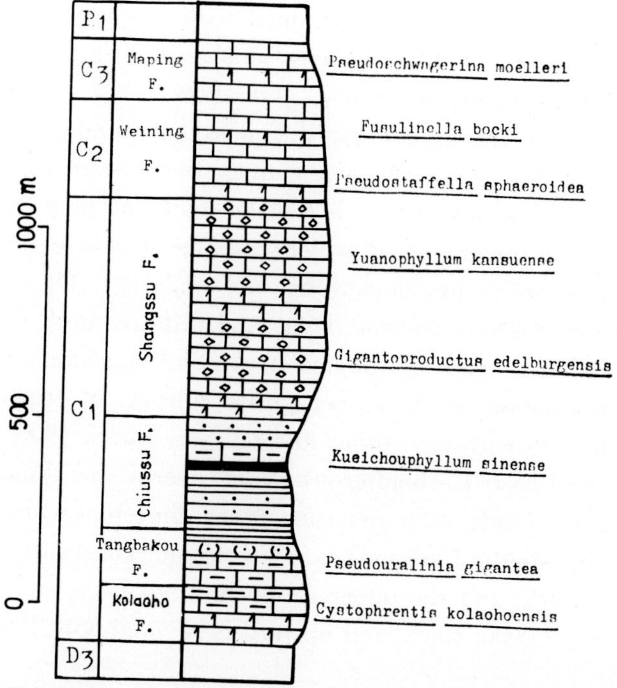

Fig. 3 Columnar section of Carboniferous strata in Dushan, Guizhou

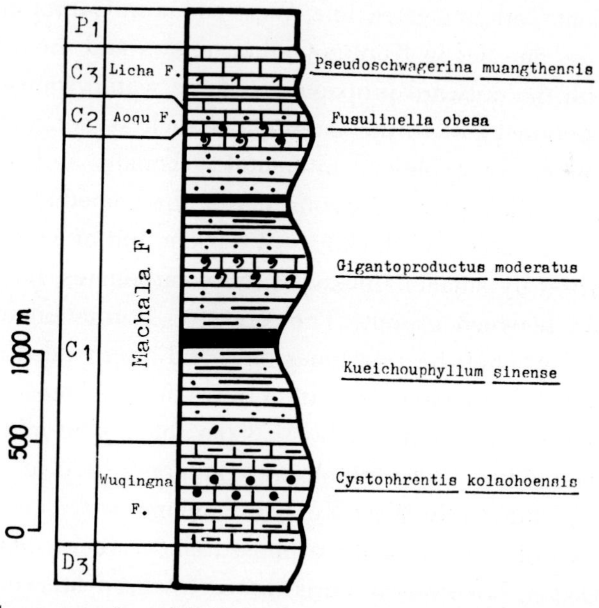

Fig. 4 Columnar section of Carboniferous strata in Qamdo, Xizang (Tibet)

5. Tianshan-Mongolia-Hinggan Ling active sedimentary region

Located in northern China, the Tianshan-Mongolia-Hinggan Ling active sedimentary region ranges from Junggar, Mts. Altay through Nei Mongol to the Northeast China. In Carboniferous period, the active sedimentation occurred in anintensively depressed geosyncline and led to a best-bedded marine strata with their petrology, facies and thickness varying on a large scale and accompanied with a great amount of volcanic lavas, porphyrite and tuffs. In Bogada Ola of East Tianshan, for example, Carboniferous strata overlie Devonian strata in unconformity, and show a lot of unconformity between formations.[1] The section shows a huge thickness and among the strata are a large amount of volcanic clastic rocks and lava with marine fossils, especially ammonites, as well as other benthos. The lower group is 6 500m thick and may be divided into two unconformitable parts, the lower one of which consists of coarse clastic rocks and limestones, and the upper are voluminous tuffs and andesites. Of the middle group, the Liushugou Formation contains volcanic lavas and tuffs of more than 3 000m thick and the Qijiagou Formation is made up of limestones and sandstones of about 900m thick. The Shirenzigou Formation of the upper group consists of seashore sandstones, conglomerates, interfingered with limestones and volcanic rocks, 260-1 200m thick. In Ulanqab, Nei Mongol, the Lower Carboniferous strata[2] overlie the Upper Devonian in unconformity and the lower group is quartz sandstones, limestones and shales, a little more than 900m thick. The Middle Carboniferous is volcanic rocks, tuffs and interbedded with limestones, 3 000m thick, and the upper group is sandstones, shales and volcanic rocks and limestones, 1 300-3 000m thick, and upward is covered with Permian strata in unconformity (Fig. 5).

6. Kunlun-Qinling and West Yunnan-Sanjiang active sedimentary region

As part of ancient Tethys geosyncline, West China in Carboniferous period had two active depocenters, i.e. one lies in Kunlun-Qinling region and the other in West Yunnan-Sanjiang region. With the outward propagation of sea water, transgressions happened to broad areas in the periphery or to the inter-uplifts such as Songpan and Ganze in West Sichuan, on which relatively stable epicontinental carbonates and clastics were deposited. In West Kunlun, the Carboniferous siliceous, clastic, basic-mediate volcanic rocks as well as carbonates reached a maximum thickness of 8 000m, but in Tewo of West Qinling, the strata are characterized by smaller thickness, more carbonates, no volcanics and the obvious unconformities between groups. The Yiwagou Formation of the lower group is dominantly composed of chert-bearing limestones and interbedded mud-limestones, 600m thick, and upward is the Lueyang Formation, a section of limestones and oolitic limestones intercalated with some sandy shales, 1 300m thick. The middle and upper groups are marine limestones, 640m thick, which are commonly divided into Minhe Formation and Gahai Formation (Fig. 6). In West Yunnan-Sanjiang active area, Carboniferous crops out in the appearance of slightly metamorphosed clastic rocks, limestones and basic extrusive rocks. In Deqen, however, it turns into crystalized limestones, tabular rocks intercalated with voluminous andesite basalts and basalts, which come to a thickness of 2 000m in all.

OUTLINE OF LITHOFACIES AND PALAEOGEOGRAPHY

The Carboniferous lithofacies and palaeogeography of China have been studied by many workers, each laying particular emphasis on them. Based on all the data available from previous investigations and in view of petroleum geology, the outline that is presented below is to focus just on petroleum-related stable sedimentary regions.

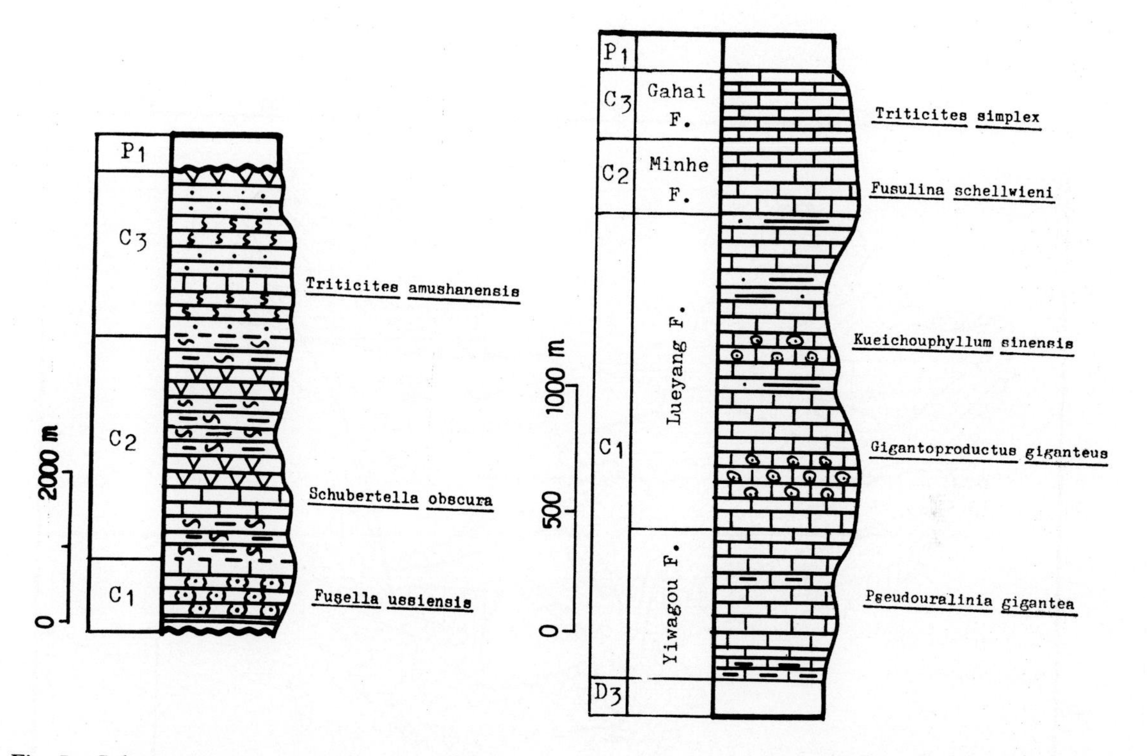

Fig. 5 Columnar section of Carboniferous strata in Ulanqab, Nei Mongol

Fig. 6 Columnar section of Carboniferous strata in Tewo, Gansu

1. Early Carboniferous

In the Early Carboniferous, North China had remained to be a continental platform since the Late Ordovician period. For the same stage, South China had undergone a regional crustal uplift starting at Late Devonian and the sea water waned backward so that the Yangtze Oldland, Kangdian Oldland, West Yunnan and West Sichuan were connected to form a broadened continent.[1] The Yangtze Platform and the Cathaysia in Southeast China were still separated by a NE-trending elongate trough through which sea water in Yunnan-Guizhou-Guangxi Sea Basin might flow through Jiangxi, Zhejiang to Lower Yangtze region. The Qinling Mountain was another trough at that time, connecting the ancient Tethys to the west and the Lower Yangtze region to the east and, because of it, the sea water transgressing to South China through the West Yunnan Trough and by way of Devonian sea to the north had spread over the Middle Tianshan Oldland and was united with the Tianshan-Mongolia-Hinggan Ling Trough of North China. Dotted in the vast water were a lot of ancient islands,[3] e.g. the Qiangtang, Litang, Batang and Qiemo Islands. On the stable platform, the gentle topography and vast sea areas provided a site for epicontinental deposition and formed facies of abundant organisms and unitary lithology. On the contrary, the topography in vast deep sea was complicated and, as a result, rowly-distributed lithology, rare living things and accompanying basic, extrusive volcanisms were obvious features of the facies (Fig. 7).

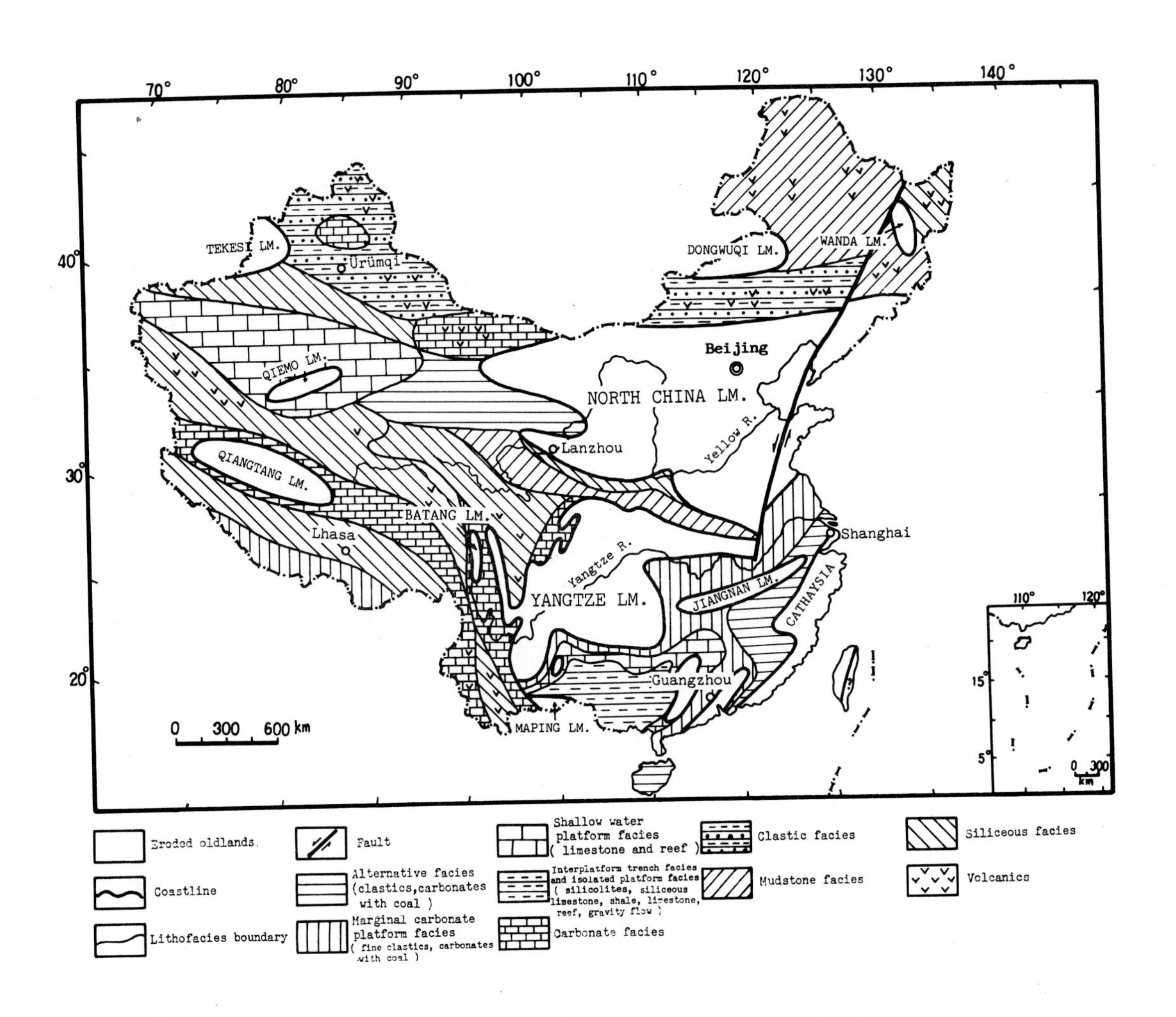

Fig. 7 Sketch map of Early Carboniferous lithofacies and palaeogeography of China

1) South China

South China at that time was a typical epicontinental sea, rich in carbonates. Facies were distributed in a regular pattern from the margin to the center of the sea, that is, the marginal carbonate platform on the margin gave way to interplatform and isolated carbonate platform towards the center.

The scope of sea water may be limited by the following boundaries:[4] the coastline along Kunming, Guiyang, through Wuhan to Cao Xian in the west and north, the Cathaysia in the southeast, the Nioushou and Kangdian Oldlands in the west and the Maping Oldland in the southwest. And in the center of the sea was the Jiangnan Oldland which ranged from Changsha to Tunxi. In addition, there were the Longmenshan Gulf and the Yanyuan-Lijiang Shallow Sea to the west of the Yangtze Oldland in which deposited sediments (Fig. 8).

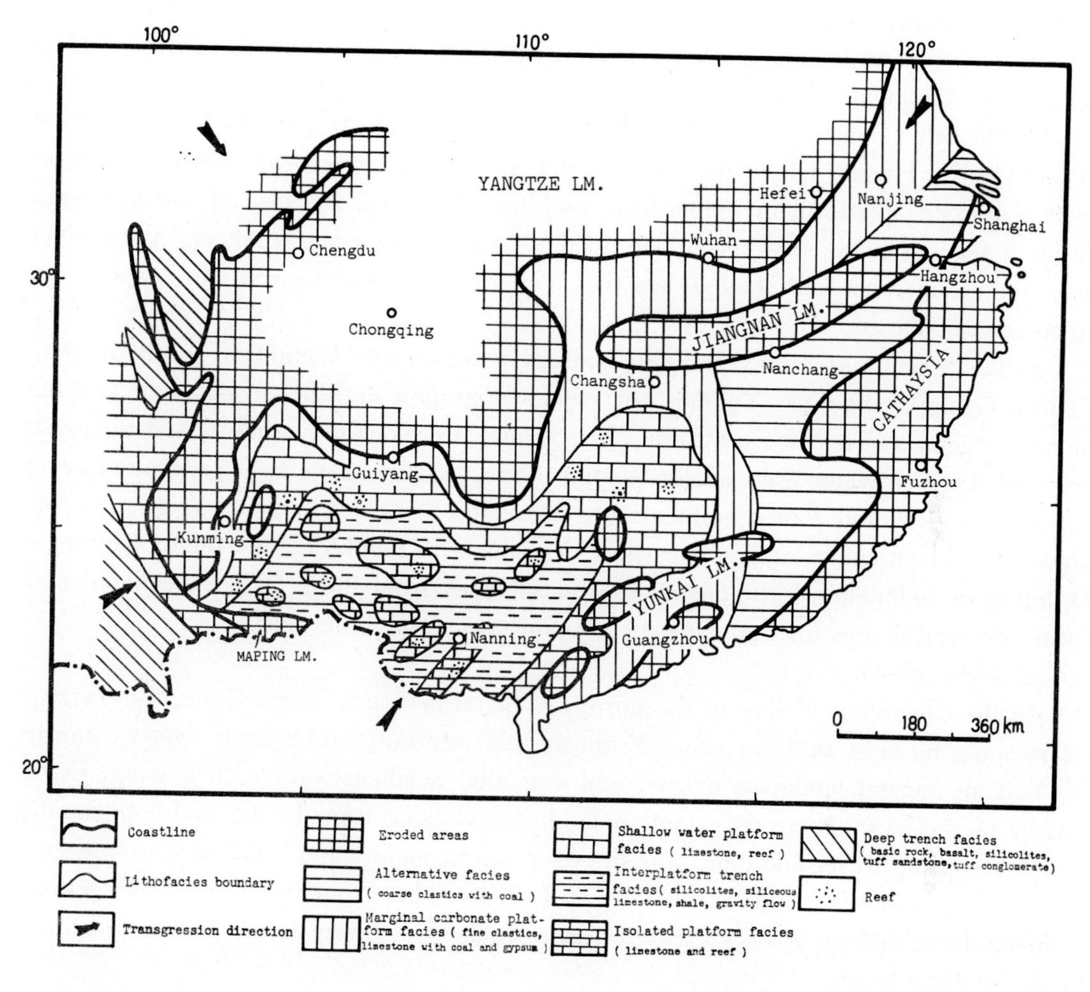

Fig. 8 Sketch map of Early Carboniferous lithofacies and palaeogeography of South China (revised from Fang Shaoxian, 1986)

Along the margin near the coastline described above developed an environment for epicontinental carbonates. The areas adjacent to oldlands are of not only facies characterized by limestones, mud limestones and sandy shales but also coal-bearing facies formed on short-lived swamp plains. In the south of Xuefeng, Jiangnan Oldland, owing to the strong erosion and surface runoff, delta sedimentary units formed both laterally and vertically may transit to epicontinental carbonate units. Taking the margin to middle of the Guizhou Oldland for example, the warm and moist climate and lush plants led to aluminium, ferriferous and coaly formations but a little far from which there became carbonate rocks. On the margin of the Maping Oldland, especially in the epiplatform depression, coastal carbonates to lagoon dolomitites are locally intercalated with gypsum or K_2O-

bearing green shales. In the meantime, the Lower Yangtze area began to depress, and the Kinling limestones deposited, which may be correlative with the Tangpakou Formation in Dushan in biota and sedimentary features, but much thinner in thickness. At the juncture of Guangdong, Hunan and Jiangxi Provinces developed marine-nonmarine transitional deposits and farther southeast to the slope zone of the Cathaysia, however, there became the dominant continental facies.

On the other hand, in the vast area of South Guizhou, East Yunnan, Central Guangxi and Central Hunan, where there were offshore basins, formed carbonates and dotted reefs within the platform or epiplatform reefs on the slope. Because of the increasing depth of sea water, the area became an open sea and the previous corals and flora which were typical in coast and shallow sea were replaced by tabular coral, brachiopods and stems of crinoids, etc.

In Luodian, Lipu, Yulin, Shanglin, Baise, Guangnan and Funing of South Guizhou, Central Guangxi and East Yunnan, the palaeogeography may be divided into two elements: troughs and isolated platforms. The former are elongately shaped and obviously oriented. The sea water was usually beyond the wave base or the light penetration, even beyond the oxidation zone. Sediments are dominated by black siliceous rocks, siliceous limestones, black shales and gravity flow deposits with planktons and minor benthoses. Different in sedimentary circumstances and schemes from these troughs, isolated platforms are typical sites for shallow sea carbonate rocks and biohermites.

2) Tarim Basin

In the Cheerchen Valley to the north of Qiemo in South Tarim Basin, the Tertiary overlies on the pre-Cambrian strata. From the unconformity and seismic data we can infer that an ancient landmass existed and extended northeastward to Ruoqiang. In the middle of the basin, there was another landmass ranging from Bachu to Luobuzhuang, which according to seismic data, was the result of an underwater uplift in Carboniferous and was covered with thin Carboniferous deposits. The rest part of the basin was an open platform and on it deposited carbonate rocks and intercalated clastics and more locally gypsum layers.

3) North China

As presented above, North China remained to be a large landmass in Early Carboniferous. However, there were two local depressions in which Lower Carboniferous strata filled. One lay at the north foot of Dabieshan Mountains and was deposited with quartz conglomerates and quartz sandstones in the lower part and quartz sandstones, Carbonaceous shales and coal layers in the upper. The other is the Gansu Corridor between the southeast platform and the north foot of Qilian Mountains. The area was filled with Early Carboniferous strata, i.e. the so-called Chounioukou Formation which consists of dominant limestones, dolomitites and sand/shales, interfingered with gypsum and coal layers.

4) Xizang (Tibet)

In the Carboniferous, Xizang could be spaciously divided into the Qiangtang Landmass in the north, the shallow sea in the south which was the northern extension of the

Indian Landmass, and the intermediate Himalayan shallow/deep sea basin.[1] Different sedimentary facies in different parts in the carbonate rocks are dominant in the north, intercalated with quartz sandstones, mud sandstones, carbonaceous shales and coal layers, while sandstones, mud limestones and shales are dominant in the south. Both parts are rich in benthos.

5) Tianshan-Mongolia-Hinggan Ling

In Carboniferous, the area was an active deep trough with a depocenter from Tianshan through Beishan to Nei Mongol. The flysch formation and intercalation of intensive volcanic extrusives are of great change in petrology, lithofacies, thickness and metamorphism, and are characterized by an intensively depressed geosyncline. Northward from the depocenter to Junggar, Mts. Altay and Hinggan Ling, transitional facies and nonmarine facies come forth. Generally, the higher the topography was northward, the more the nonmarine facies sediments developed. The Wanda Mountains was an ancient landmass and was eroded to Proterozoic Erathem. In Junggar the Lower Carboniferous strata was marine deposits with a medium thickness, but the sea water retreated in Late Carboniferous and transitional facies were intercalated with coal layers.

2. Middle to Late Carboniferous

As the largest one since Late Palaeozoic, the transgression in Middle to Late Carboniferous ended the continental history of North China since Late Ordovician. In the periphery of the platform remained some uplifts, e.g. NNE-trending Jiaoliao Landmass in the east, the EW-trending Yinshan Landmass in the north and North Qinling-Dabieshan Landmass in the south, the latter being the protective screen between the North China Epicontinental Sea and the Qinling Trough.[3] In South China, the sea water in West Hubei Gulf, Longmenshan Gulf and Yanyuan-Lijiang shallow sea started to transgress in Early Carboniferous and continued to spread so widely that the scope of the Yangtze Landmass decreased significantly. It was also owing to the southeastward spreading of South China Sea that the Cathaysia was reduced to the recent coast area, east of Shaoxing, Zhenghe and Haifeng. The warm climate, slow crustal depression, numerous organisms and relatively gentle palaeogeography in the above seas had grounded a typical epicontinental sedimentation with unitary petrology. At the end of Late Carboniferous, both North China and South China began with a stage of regional rising-up and almost became a united continent, except for a few intensively depressed areas representing disharmonious crustal movement (Fig. 9).

Apart from North and South China, other areas of China still retained the palaeogeography of Early Carboniferous.

1) North China

From Middle to Late Carboniferous, the whole North China Landmass subsided. Separated by the Wulangeer-Wuqi Landmass, there were two transgressing seas. One came from the Qilian Trough and extended far eastward to Zhuozishan. Under the circumstances of complicated gulf, lagoon or shallow/coast sea, the Tsingyuan, Yanghukou and Taiyuan Formations, which are composed of dominant clastics and carbonates and thin layers of interfingered coal and siderite, were deposited. The other flow was from the Taizihe River

Basin in Northeast China, where the thickness of marine limestones and coal beds of the Penchi Formation may reach 300m. But the strata get thinner and thinner from Kaiping in Hebei to Taiyuan in Shanxi, even to nothing in some places. The Penchi Formation can not be found any more to the south of Xinxiang-Puyang and to the west of the Yellow River. From the facies changes we come to the conclusion that the sea water was derived from the Taizihe River Basin and from it transgressed to the whole North China. In Late Carboniferous, the sea water spread further and the south part of the Wulangeer-Wuqi Landmass was transgressed by the emerged sea water from east to west. Because the well data reflect an increasing thickness and amount of marine strata from Ertuokeqi through Shanxi to South Anhui, it seems reasonable to infer that North China was a coast or shallow sea in Late Carboniferous and that the sea water was derived from the southeast (Fig. 10).

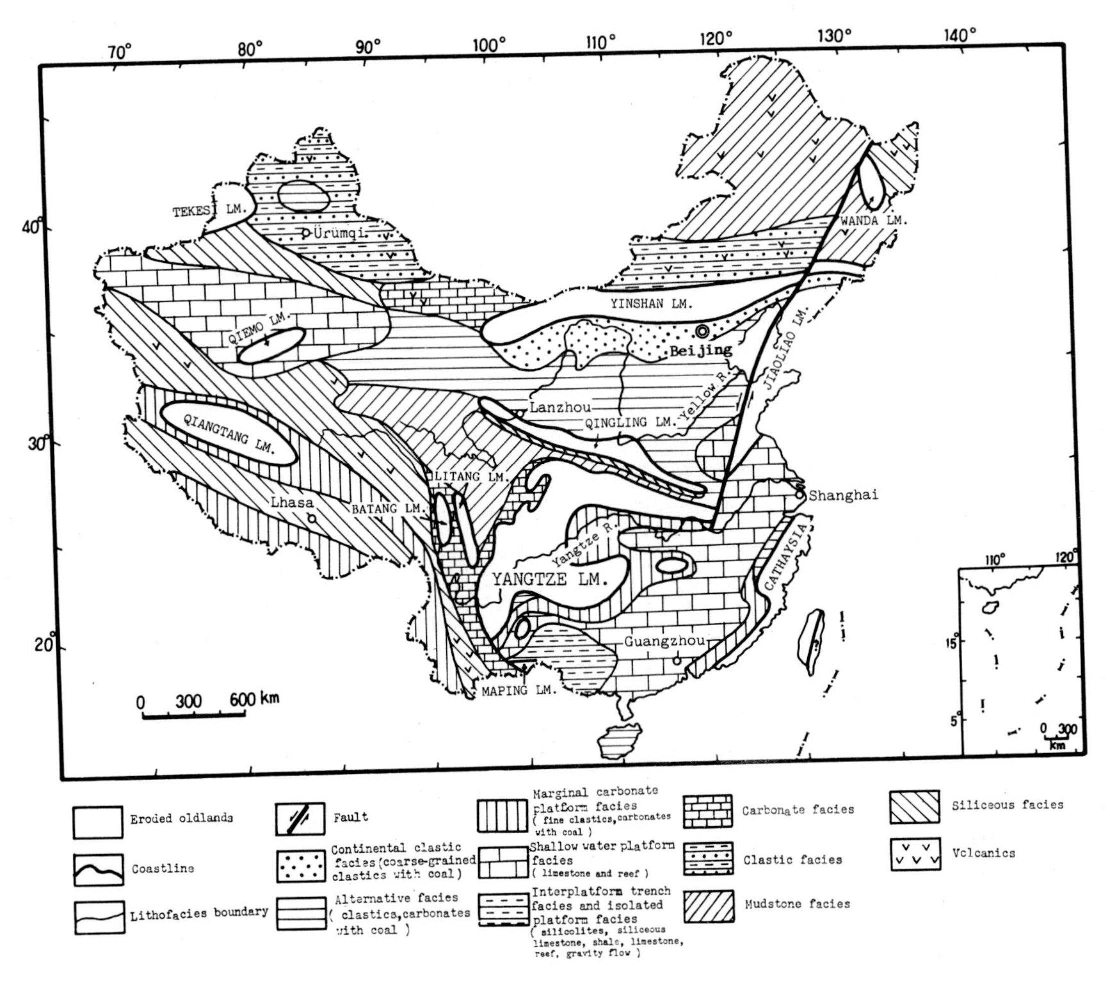

Fig. 9 Sketch map of Middle to Late Carboniferous lithofacies and palaeogeography of China

2) South China

South China was transgressed on an unprecedented scale in Middle to Late Carboniferous and the scope of continental erosion areas was notably reduced. There was coastal clastic sedimentation of short duration near the hilly area in Zhejiang, Fujian and Guangdong. It was usually tidal flat of marginal sea near the coastal area of Xuefeng, Upper Yangtze and Kangdian Landmasses. Because of the dry and hot climate and ill developed surface current, gypsum and penecontemporaneous dolostone formed in the tidal flat and intertidal area. The rest broad area was an open platform where shallow sea carbonate facies is dominant, and the thickness of limestones varies slightly. In accordance with the change of water depth and energy, the tidal flat formed. However, isolated platform and interplatform troughs were developed in West Yunnan, South Guizhou and Guangxi. On account of deeper water, the troughs were filled with thick strata of limestones with siliceous sediments such as chert masses and stripes, and with rare fossils. They had facies of trough sediments. On isolated platform, however, shallow water limestone and reef facies developed (Fig. 11).

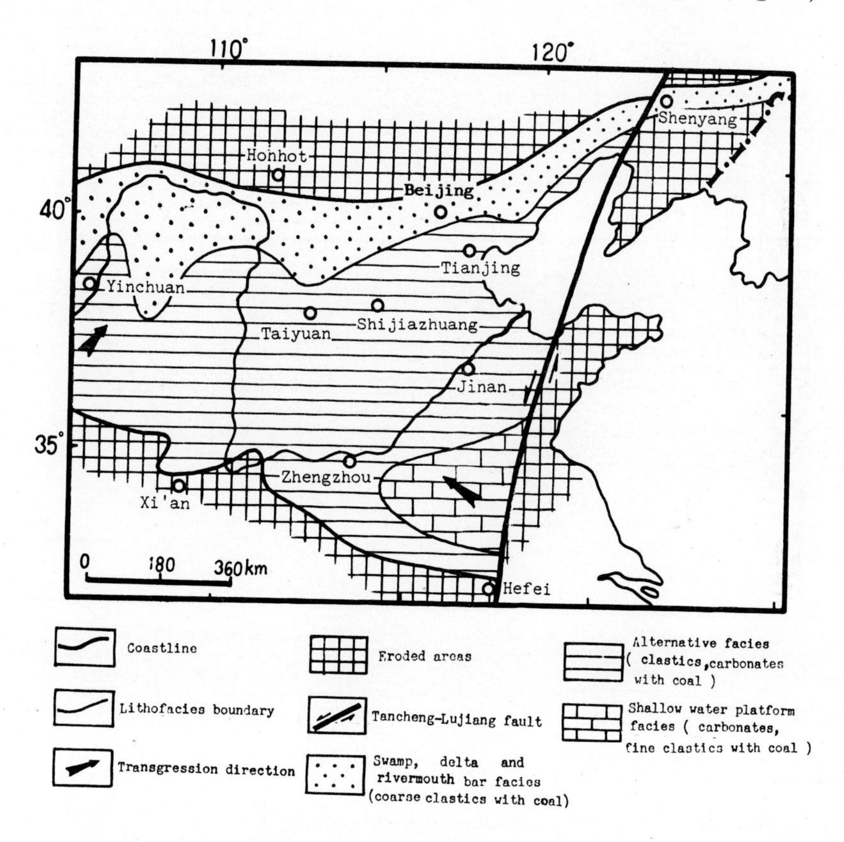

Fig. 10 Sketch map of Late Carboniferous lithofacies and palaeogeography of North China

3) Kunlun-Qinling and West Yunnan-Sanjiang

Characteristic of intensive crustal activity, basically geosynclinal sedimentation, occurred in the deep or shallow sea basin of the area in Carboniferous. However, the differentiation within the geosyncline in the intensity of depression, topography and volcanism was so great that a variety of sediment types with obvious change in thickness are formed. Generally speaking, the area is divided into three types of sedimentary circumstance: (1) the relatively stable mudstone-limestone-dolomite facies type, with minor thickness and abundant fossils; (2) the transitional clastics-limestone facies type in shallow trough basin, with moderate thickness and carrying benthos and planktons; and (3) the intensively active basic rocks-extrusive rocks-clastic rocks facies type, with huge thickness and rare fossils. The first type is seen in Tengchong, Lincang-Zhengkang, Ailaoshan-Jinping and Batang-Zhongdian, the transitional shallow basin type is developed in Songpan-Li Xian while the active type is typical in West Kunlun, North Qinling, Lanping-Simao and Baoshan-Shidian, etc.

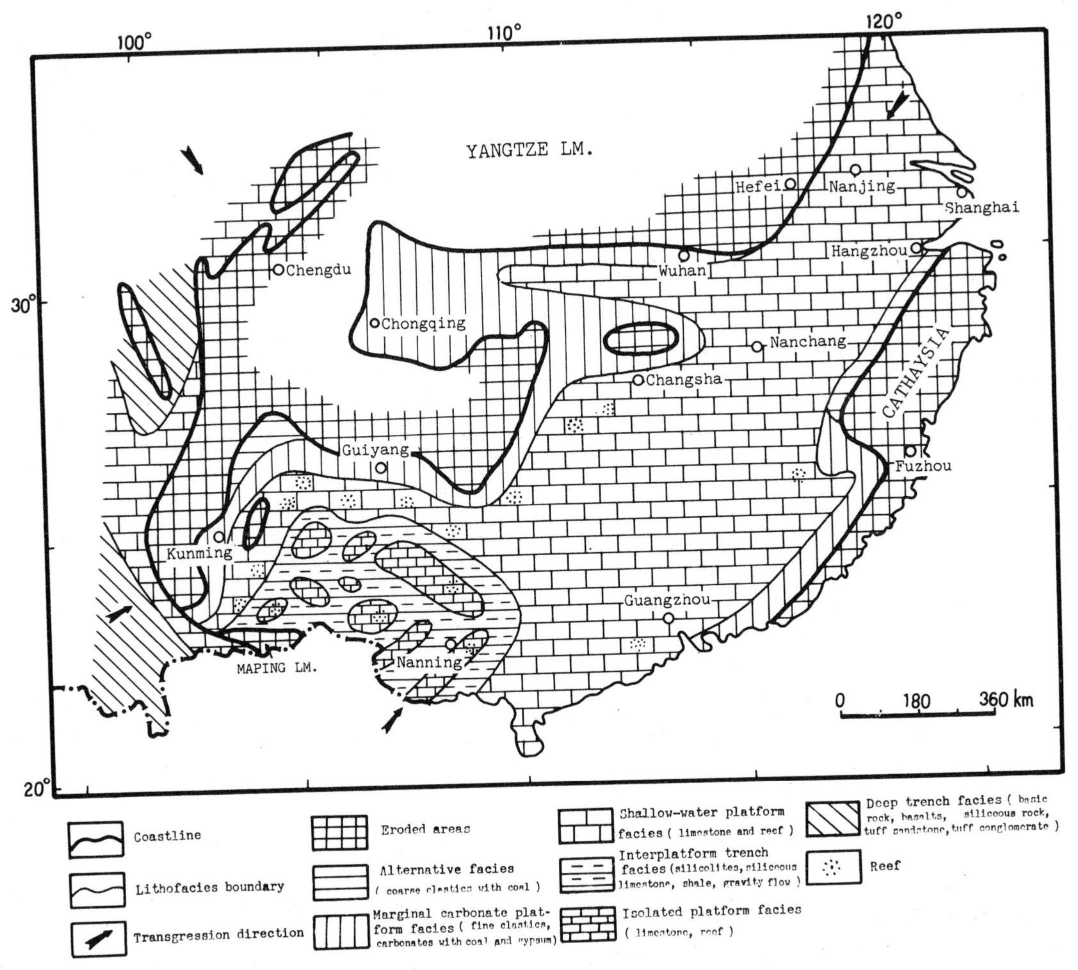

Fig. 11 Sketch map of Middle to Late Carboniferous lithofacies and palaeogeography of South China (revised from Fang Shaoxian, 1986)

PETROLEUM POTENTIAL

It is considered that oil and gas-bearing basins on the earth are distributed under the control of the tectonics, petrology and palaeogeography, and so are their types and petroleum potential. As a result, it is a foundational and important work for petroleum explorers to conduct reseaches on the petrology and palaeogeography of basins. It is essentially helpful for us to make known the properties and scope of basins, position and facies group favorable for petroleum generation and preservation as well as how the covering strata are developed. Moreover, it can provide us with a clue to classify the possible traps and find the law of their distribution.

To do so, we have to look for geological conditions on which petroleum reservoirs were generated in basins. In a stable negative tectonic units as in Tarim, South China and North China stable areas, if the depressions or faulted block depressions within or on the margin of the basins depressed constantly for a long time and sediments were filled to a sufficient thickness with abundant organic matters, this area may be listed as a potential one for oil and gas.

The petroleum potential of Carboniferous stable sedimentary regions of China has been investigated for ages, in aspects of regional tectonics, sedimentary basins, facies, palaeogeography, organic geochemistry, geohistory and geothermal evolution and so on. Geophysical prospecting, exploring wells and the gas flow in some areas in recent years have added a new understanding to Carboniferous petroleum potential.

1. Favourable generation environments

By means of the study of tectonics and sedimentary facies, we think the following types of Carboniferous basins are of good petroleum potential:

1) Closed depressions in large-scale basins within platform, as the Manjiaer Depression and Hetan Depression in Tarim, the south and middle parts of North China;

2) Epicontinental basins on the platform slope, as the transitional zone between the Yangtze-Xuefeng Landmass and the Yunnan-Guizhou-Guangxi Sea Basin;

3) Extentive fault-block basins, as the interplatform troughs in South Guizhou, Central Guangxi and East Yunnan;

4) Epicontinental trough basins on the margin of platforms, as in South and West Guizhou Yanyuan-Lijiang Epicontinental Sea, Longmenshan Gulf and the west margin of North China Platform. Composed of black shales, fine mudstones, organic mudstones and muddy limestones as well, the sedimentary facies in these negative structure units are all favourable for the accumulation of organic sources.

2. Assemblage of source-reservior-cap rocks

In view of strata in all of petroleum provinces, it can be found out that platform and beach reefs and dolomitized rocks which usually developed in the platform margin and closed areas within platform are usually good reservior rocks. In addition, subsidiary pores formed in the diagenesis of carbonate rocks also provided good space for oil accumulation.

In view of the distribution of sedimentary environments and facies, North China and the Tarim Basin have better assemblages of the source-reservior-cap rocks than South China because the vibration movement of the crust and changes in source rocks and water dynamics led to a polycycled sedimentary sequence in North China. On the other hand, the carbonate-dominated South China has better conditions for lateral oil migration than North China and the Tarim Basin.

3. Traps

The analysis that follows will lay emphasis on traps for several kinds of non-structural pools.

1) Unconformity reservior traps

Between the Carboniferous and the underlain Devonian was a regional degradation unconformity and was also a major passage way for petroleum migration. Because the unconformity traps are of properties of broad extension and good preservation they are worth paying great attention. Some findings of natural gas in the Sichuan Basin may have something to do with it.

2) Reef reservior traps

Organic reefs were largely developed on the carbonate platform and on its marginal slope, known as the orientation of migration and place of accumulation for hydrocarbons. The existence of asphalt veins and oil shows in exposed reef rocks indicates that hydrocarbonic accumulation processed in them in the geohistory.

3) Reservior traps in isolated platforms

Well-developed isolated carbonate platform in South Guizhou, West Yunnan and Central Guangxi provide good conditions for oil accumulation for the shield of low-permeability strata around them. Moreover, they were surrounded by interplatform troughs which were filled with plentiful source rocks, therefore, it is very possible to form reservior traps on the platforms.

4) Reservior traps of clastic rocks by gravity flow

In deep water on the platform slopes, clastic rocks by gravity flows were developed not only as good source rocks but also as excellent traps and preservation for oil and gas. The outcrops of asphalt veins filled in exposed rocks of the type indicate the trace of petroleum migration in the past.

4. Preservation

The preservation of oil and gas turns out to be another important fact in resource exploration. If the Carboniferous strata were overlain by later strata, for example, the Meso-Cenozoic continental deposits, the area can be listed as a well-preserved one for Carboniferous petroleum. Among the areas are the Tarim Basin, the North China Basin, the Ordos Basin, the Gansu Corridor, the Sichuan Basin, the Jianghan Basin and the Lower Yangtze region.

REFERENCES

1. Wang Hongzhen and Liu Benpei, In: A Course in Geological History, Geological Publishing House (1980), 179.
2. Yang Jingzhi, Wu Wangshi et al, In: Stratigraphic Correlation Chart in China with Exploration Text, Science Press (1982), 124.
3. Guan Shicong et al, In: The Change of Land and Sea, Marine Sedimentary Facies and Hydrocarbons in China (Late Proterozoic to Triassic Ages), Science Press (1984). 68.
4. Huang Jixiang et al, In: Publication of the Southwest Petroleum College, Sichuan, 1986.

XIe Congrès International de Stratigraphie et de Géologie du Carbonifère Beijing 1987,
Compte Rendu 1 (1991): 157-168

PROJECT OF AN INTERNATIONAL SCIENTIFIC CLASSIFICATION OF SOLID FOSSIL FUELS

B. Alpern
(23 bis, rue des Cordelieres, 75013 Paris, France)
M. J. Lemos de Sousa
(University of Porto, Praça de Gomes Teixeira 4000 Porto, Portugal)

ABSTRACT

The scientific classification chart presented concerns coals (humic and sapropelic), mixtes and oil shales.

This project is in the process of being discussed for 8 years in the ICCP. Many points and basic principles have been accepted, the chart is already used in some countries, and it has been presented to the United Nations in Geneva for examination. The classification is based on 3 main parameters:

RANK or degree of coalification, is the result of palaeotemperature-time effects. It is measured by vitrinite reflectance, if it is greater than 0.6%, and moisture-holding capacity if it is lower. The coalification scale is divided in nine steps by means of 3 main words: LIGNITE, BITUMINOUS, and ANTHRACITE, each class being subdivided into 3 subclasses by using the prefixes: hypo, meso, meta.

TYPE, which is related to transitional conditions, aerobic to anaerobic, dry to wet. It is expressed by 3 words: FUSIC, VITRIC, LIPTIC, from continental to brachish marine deposits and quantified by group-maceral percentages.

FACIES, which varies from coaly to shaly and which is related to the quantity and intimacy of mineral admixtures. It is measured by ash percentage and the washability potential, as expressed by the fraction with less than 10% ash. Three words are used: COAL, MIXTE, SHALE.

In conclusion, the system uses words, which convey a clear meaning to a majority of users without necessity for translation; also the three parameters can be graphically represented on the chart.

This scientific classification is conceived mainly for geological purposes: basin characteristics, reserve calculations, and *in situ* evaluation. However, it can also serve other practical applications such as carbonization and beneficiation. We intend, not to replace but to correlate national systems. This chart serves an educational purpose, being able to play a role in teaching and scientific intercourse; purposes which cannot be attained with the coded number system elaborated independently in the United Nations for trade and technical use.

INTRODUCTION

1. Previous works

The classification presented here is the emended version of a chart, the basic principles and parameters of which were published by Alpern [1,2,3] and by Lemos de Sousa.[8] This chart has been discussed within the ICCP and has been largely accepted after a larger number of enquiries and votes (internal ICCP reports).

This classification is independent of the system of codification for trade and technical use which has been elaborated in 1987 by the United Nations in Geneva for medium and high rank coals, and which uses a 14-digit code number based on 8 parameters.[7] However, it has some of the parameters in common with coding system, viz. reflectance, maceral composition, and ash content (Table 1); also the boundary between low and medium rank coals is placed at the same reflectance value.

The present chart, using words instead of code numbers, has been selected by the French Government and also presented to the United Nations in Geneva to classify on a scientific basis all coals from low to high rank.[9]

This so-called "Alpern System" is being used already to classify coal basins in South Africa,[6] and in Brazil.[4]

A Working Group on Classification is presently active within ICCP under the leadership of Professor Lemos de Sousa. The latest report presented to ICCP uses this chart to give results corresponding to 33 basins in 6 countries both from Laurasia and Gondwanaland.

2. Area covered by the classification

The classification embraces all carbonaceous rocks which are solid fossil fuels.

It includes: humic and sapropelic coals;
mixtes, humic mixtes and oil shales.

It excludes: peats, because some peats do not qualify as fossil fuel;
graphite, because it is a mineral;
solid bitumens, because they are related to hydrocarbons and are often not generated *in situ*.

The materials classified must be representative of individual seams, and exclude industrial blends.

3. Means of the classification

The classification is based on:

simple descriptive words generally understandable without translation;

a small number of ranked parameters used successively and never in competition;

a graphic presentation which facilitates comparisons and teaching.

4. Purpose of the classification

This classification is basically scientific and geological but can also be for other application such as:

Table 1 Codification system for medium and high rank coals. United Nations, Geneva, 1987

Basic parameters	Mean random vitrinite reflectance %		Reflectogram(*)		Maceral group composition Vo1.%(mmf) 4-Inertinite; 5-Liptinite				Free swelling Index		Valatile matter content mass % (daf)		Ash content mass % (d)		Total sulphur content mass % (d)		Gross Cal. Val. MJ/kg (daf)	
Digits	1;2		3		4	Inertinite	5	Liptinite	6		7;8		9;10		11;12		13;14	
Code no.	02	0.20-0.29	0		0	0-<10	0	exempt	0	0-1/2	48	≥48	00	0-<1	00	0.0-<0.1	21	<22
	03	0.30-0.39	1		1	10-<20	1	>0-<5	1	1-11/2	46	46-<48	01	1-<2	01	0.1-<0.2	22	22-<23
	04	0.40-0.49	2		2	20-<30	2	5-<10	2	2-21/2	44	44-<46	02	2-<3	02	0.2-<0.3	23	23-<24
	05	0.50-0.59	3		3	30-<40	3	10-<15	3	3-31/2	42	42-<44	03	3-<4	03	0.3-<0.4	24	24-<25
	06	0.60-0.69	4		4	40-<50	4	15-<20	4	4-41/2	40	40-<42	04	4-<5	04	0.4-<0.5	25	25-<26
	07	0.70-0.79	5		5	50-<60	5	20-<25	5	5-51/2	38	38-<40	05	5-<6	05	0.5-<0.6	26	26-<27
	08	0.80-0.89			6	60-<70	6	25-<30	6	6-61/2	36	36-<38	06	6-<7	06	0.6-<0.7	27	27-<28
	09	0.90-0.99			7	70-<80	7	30-<35	7	7-71/2	34	34-<36	07	7-<8	07	0.7-<0.8	28	28-<29
	10	1.00-1.09			8	80-<90	8	35-<40	8	8-81/2	32	32-<34	08	8-<9	08	0.8-<0.9	29	29-<30
	11	1.10-1.19			9	≥90	9	≥40	9	9-91/2	30	30-<32	09	9-<10	09	0.9-<1.0	30	30-<31
	12	1.20-1.29									28	28-<30	10	10-<11	10	1.0-<1.1	31	31-<32
	13	1.30-1.39									26	26-<28	11	11-<12	11	1.1-<1.2	32	32-<33
	14	1.40-1.49									24	24-<26	12	12-<13	12	1.2-<1.3	33	33-<34
	15	1.50-1.59									22	22-<24	13	13-<14	13	1.3-<1.4	34	34-<35
	16	1.60-1.69									20	20-<22	14	14-<15	14	1.4-<1.5	35	35-<36
	17	1.70-1.79									18	18-<20	15	15-<16	15	1.5-<1.6	36	36-<37
	18	1.80-1.89									16	16-<18	16	16-<17	16	1.6-<1.7	37	37-<38
	19	1.90-1.99									14	14-<16	17	17-<18	17	1.7-<1.8	38	38-<39
	20	2.00-2.09									12	12-<14	18	18-<19	18	1.8-<1.9	39	≥39
	21	2.10-2.19									10	10-<12	19	19-<20	19	1.9-<2.0		
	22	2.20-2.29									09	9-<10	20	20-<21	20	2.0-<2.1		
	23	2.30-2.39									08	8-<9			21	2.1-<2.2		
	24	2.40-2.49									07	7-<8			22	2.2-<2.3		
	25	2.50-2.59									06	6-<7			23	2.3-<2.4		
	26	2.60-2.69									05	5-<6			24	2.4-<2.5		
	27	2.70-2.79									04	4-<5			25	2.5-<2.6		
	28	2.80-2.89									03	3-<4			26	2.6-<2.7		
	29	2.90-2.99									02	2-<3			27	2.7-<2.8		
	30	3.00-3.09									01	1-<2			28	2.8-<2.9		
	31	3.10-3.19													29	2.9-<3.0		
	32	3.20-3.29													30	3.0-<3.1		
	33	3.30-3.39																
	34	3.40-3.49																
	35	3.50-3.59																
	36	3.60-3.69																
	37	3.70-3.79																
	38	3.80-3.89																
	39	3.90-3.99																
	40	4.00-4.09																
	41	4.10-4.19																
	42	4.20-4.29																
	43	4.30-4.39																
	44	4.40-4.49																
	45	4.50-4.59																
	46	4.60-4.69																
	47	4.70-4.79																
	48	4.80-4.89																
	49	4.90-4.99																
	50	≥5.00																

(*) Reflectogram characteristics

Code	Standard deviation			Type
0	≤ 0.1			seam coal
1	> 0.1	≤ 0.2	no gap	seam coal
2	> 0.2		no gap	simple blend
3			no gap	complex blend or anthracite
4			1 gaps	seam coal
5		more than	2 gaps	blend with 1 gap
			2 gaps	blend with 2 gaps
				blend with more than 2 gaps

Example for a good coking coal:

12 2 04 9 22 05 01 36

the exploration and the characterisation of basins;
the correlation between national systems;
the calculation of reserves and resources using a common basis;
seam correlation and coal geology in general;
the reconstruction of palaeoprovinces;
in situ gasification;
carbonization, etc.

CLASSIFICATION PARAMETERS

Three main parameters have been selected (Fig. 1): RANK for coalification; TYPE for petrographic constitution; FACIES for ash content and washability potential.

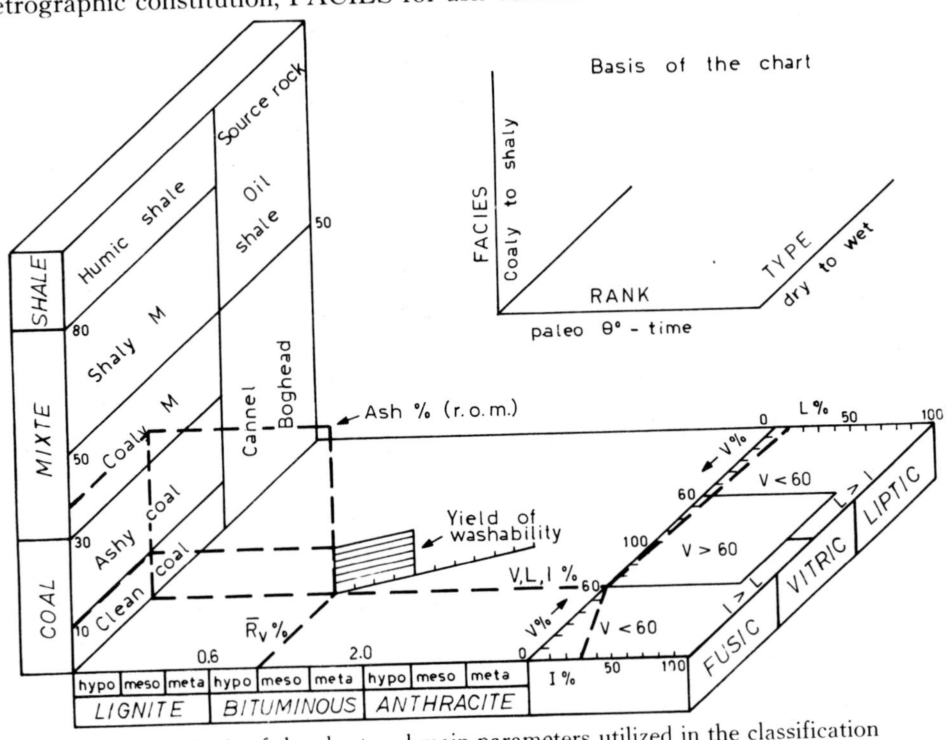

Fig. 1 Basis of the chart and main parameters utilized in the classification

1. Rank (Fig.2)

1) This is related to the palaeotemperature-time effect as represented by coalification. Three main steps are recognised: LIGNITE, BITUMINOUS, ANTHRACITE; these correspond the low, medium and high rank coals respectively. The parameter for this ternary division is the mean random reflectance of vitrinite using the boundary values 0.6% and 2.0% (Fig.2). Names such as BITUMINOUS and ANTHRACITE are largely accepted, and "LIGNITE" is preferred to "BROWN COAL" because part of the low rank coals are black; "lignite" does not imply a specific colour. A supplementary division is obtained for rank by using the prefixes, "hypo", "meso", and "meta" for each of the 3 main categories (Fig.2).

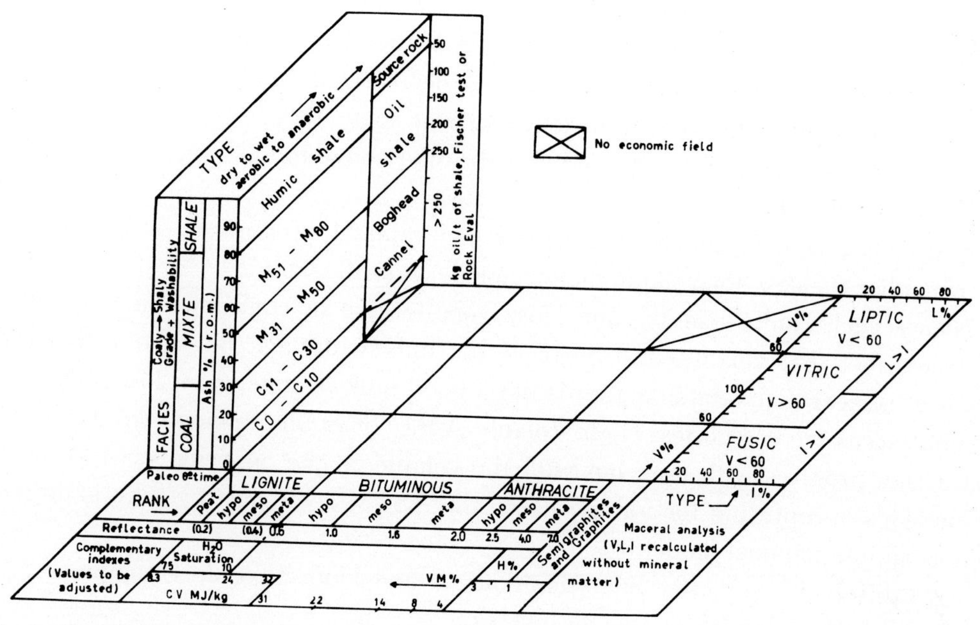

Fig. 2 Detailed chart for the so-called "Alpern's System of Classification of Solid Fossil Fuels"

2) In the United Nations at Geneva, the Group of Experts on the Utilization and Preparation of Solid Fuels has adopted a partly similar but more complex definition for the limit between low and medium rank coals using random vitrinite reflectance ($\bar{R}_v$) and calorific value on a moist, ash free basis (CV_{maf}).

Low rank coals: $CV_{maf} < 24$ MJ/kg and $\bar{R}_v < 0.6\%$

Medium and high rank coals: $CV_{maf} \geqslant 24$ MJ/kg

or

$CV_{maf} < 24$ MJ/kg but $\bar{R}_v \geqslant 0.6\%$.

3) To respect our principle of hierarchy and non-competition between parameters we have retained only vitrinite reflectance in our chart; the calorific value is subject to many errors (and, consequently, to correction factors) due to mineral matter being transformed during heating, the overall effect being generally endothermic.[5] This is mainly valuable for Mixtes (ash % between 30 and 80) since the effect of mineral decomposition on thermal balance depends on the kind of the minerals (clay, carbonates, sulfides, quartz, etc.).

4) For coals with $\bar{R}_v < 0.6\%$ we propose to use moisture-holding capacity (30^{0}C, 96% humidity) as a supplement to reflectance, since this varies from about 10 to 75%. However, for these categories calorific value is a supplementary parameter of utmost practical importance; it varies from 14.6 to 24.4 MJ/kg.[10]

The ternary subdivision of lignites in hypo, meso and meta, using specific indexes such as moisture-holding capacity and calorific value will be elaborated in detail. For Humic Mixtes and Oil Shales important correction factors will have to be introduced in order to calculate the real value of these indexes on a pure coal basis. In this case the reflectance value of huminite macerals will remain important and helpful.

5) Regarding the limit between BITUMINOUS and ANTHRACITE the reflectance value of 2.0% corresponds to the limit of coking properties, even at a very high speed of heating. It also corresponds to the more rapid decrease of the H/C values in the Van Krevelen diagram. In fact a transition zone exists at about 1.9%.

6) The advantages of the proposed classification above the US one[10] and related systems for rank are that:

Reflectance is independent of maceral and mineral proportions and composition. This is neither the case for volatile matter nor the calorific value.

Names such "high volatile" and "low volatile" are in contradiction with the well established fact that volatiles can decrease not only with rank but also with inertinite content. It is clear that volatile matter is not a good rank parameter for low rank coals.

The successive categories: High Volatile A,B,C or Subbituminous A,B,C in the US chart do not progress with rank but with the volatile, in the opposite sense to the coalification. This is confusing for foreign users; the ASTM coal chart by rank is difficult to adopt as an international model.

Two different categories: Subbituminous A and High Volatile Bituminous C have the same calorific value limits: 24.4-26.7 MJ/kg.

7) The other complementary specific indexes to be elaborated in detail for a ternary subdivision of the 3 main rank categories are:

Dilatation, measured either by the Free Swelling Index (FSI) or by the dilatometer test for BITUMINOUS coals.

Volatile matter which can be reintroduced as a rank parameter when the coalification jump of liptinites is achieved (higher than $\bar{R}_v = 1.5\%$).

Hydrogen content which is particularly valuable near the end of coalification with a 1% limit value between Meta-anthracites and Semigraphites. The end of the domain to be classified could be situated at H=0.5%, Random reflectance=11%. Beyond these values begins the mineralogical domain of graphite.

2. Type

1) We have tried to use properties related to palaeodepositional conditions: dry to wet, aerobic to anaerobic, resulting in fusinisation to gelification processes of organic matter decomposition.

The 3 names: FUSIC, VITRIC, LIPTIC are clear without translation and they do not imply a specific behaviour as for "intertinite".

The quantitative limits are given in the chart (Fig. 2):

VITRIC: Vitrinit > 60%;

FUSIC: Vitrinite < 60%, Inertinite>Liptinite;

LIPTIC: Vitrinite < 60%, Liptinite<Inertinite.

2) Maceral analysis is to be done on:

representative samples of raw coal seams;

washed fractions (ash< 10%), whenever possible, so as to agree with the ISO standard.

In any case the maceral analysis is finally expressed and plotted on the chart on a pure organic matter basis but it is necessary to specify whether maceral analysis has been done on a raw or on a washed product because the results are different.

3. Facies

1) This parameter is related to the proportion and intimacy of mineral admixtures resulting from variations between coaly and shaly facies (or aspect). The FACIES concept differs from "Grade" because it includes not only the ash content but also the overall washability potential. We believe that it is this last property which permits to distinguish between transportable-marketable clean coals and non-washables ones which can only be used locally. Some basins do not provide any economic fraction with less than 10% of ash, which is the quality required for carbonization, chemical analysis, etc. On the other hand, *in situ* evaluation implies the use of coal as it is, without extraction or washing. Nevertheless, even in the case of *in situ* gasification, the mode of distribution of mineral matter, either intimate or in discrete beds, is also important.

2) The washability potential in terms of yield of washability is calculated either by a laboratory test or in industrial plants. During exploration often only drill-cores or column samples are available. In this case the laboratory tests may be applied to a representative sample of 1kg crushed to a size below 1mm. A washability curve using successive density fractions is then to be established and the fraction with less than 10% ash is calculated for plotting in the chart. Such a test is not intended to predict the economic performance of a future washery plant but just to express the overall potential.

For certain categories, such as Oil Shales and Hypolignites, which are not usually washed, the washability potential does not need to be established.

3) The introduction of a parameter related to the quantity and mode of occurrence of mineral matter is supported by the fact that mineral transformations have negative effects on the calorific value (except pyrite) and a diluting effect on volatiles formed during gasification and combustion. Moreover, Mixtes and even Shales have an industrial application, such as bricks and expanded shales. Oil Shales have also great industrial potential; they are used in China and the USSR and will be used in many other countries when it becomes necessary. These carbonaceous rocks which have negligible energetic potential at present, will be then relevant to a classification covering all solid fossil fuels.

4) The following categories have been proposed for FACIES:

(1) Humic: VITRIC and FUSIC types.

COAL—ash <30%

Clean Coal—ash < 10%

Ashy Coal—ash 11-30%

MIXTE—ash 31-80%

Coaly Mixte—ash 31-50%

Shaly Mixte—ash 51-80%

SHALE (Humic)—ash > 80%

(2) Sapropelic: LIPTIC type.

Sapropelic coals such as Cannel Coals (with spores) and Bogheads (with alga) are generally neither clean nor easily washable. Their ash content should be lower than 50%.

Higher than 50% ash content we have Oil Shales which should rather be classified with reference to their oil yield during pyrolysis (Fisher test, Rock Eval or similar ones). The proposed boundaries for Oil Shales are: 50 to 250kg oil/t of shale. The limit between Oil Shale and Sapropelic Coals: 250kg oil/t corresponds approximately to 50% of organic matter of liptic type with a minimal conversion factor to oil of 50%. Source rocks are carbonaceous rocks of liptic type with less than 10% of organic matter corresponding to an oil yield of less than 50kg/t.

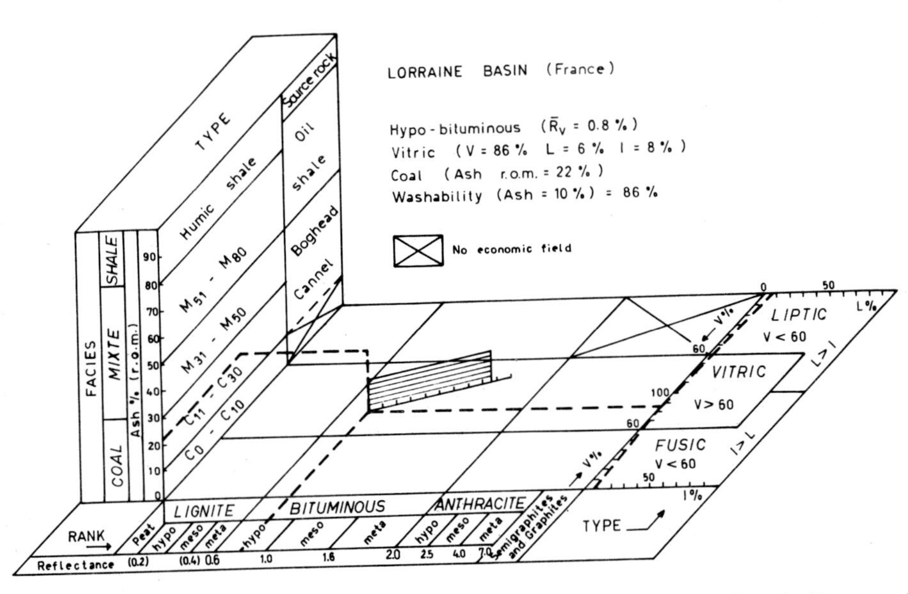

Fig. 3 Example of a coal from the Lorraine Basin (France) plotted in the Alpern's System, detailed chart

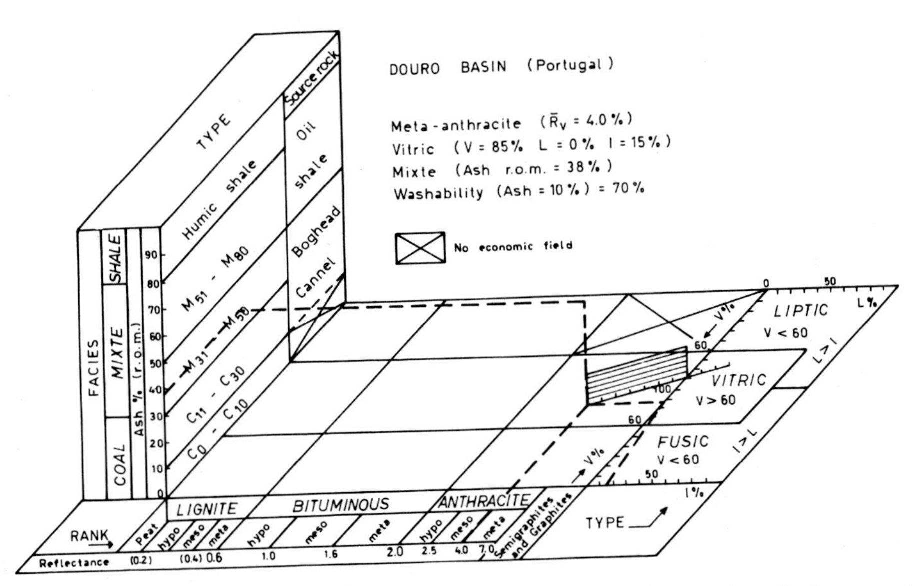

Fig. 4 Example of a coal from the Douro Basin (Portugal) plotted in the Alpern's System, detailed chart

GRAPHIC REPRESENTATION AND EXAMPLES

1) A detailed chart is presented as Fig. 2.

Some specific examples are given for the following basins:

Fig. 3—Lorraine (France);

Fig. 4—Douro (Portugal);

Fig. 5—Candiota (Brazil);

Fig. 6—Agadès (Niger).

2) The possibility is given to plot the maceral constitution not on 3 orthogonal axes as in Fig. 2 but in a classical triangular diagram. In this case supplementary space is available for transitional categories in the centre of the triangle.

Examples are given for the northwestern part of the Slovenia Basin, Yugoslavia with both graphical presentations (Figs. 7 and 8).

3) For clean coals the vertical ash axis becomes available for another property. An example is given where the Free Swelling Index (FSI) is used (Fig. 9).

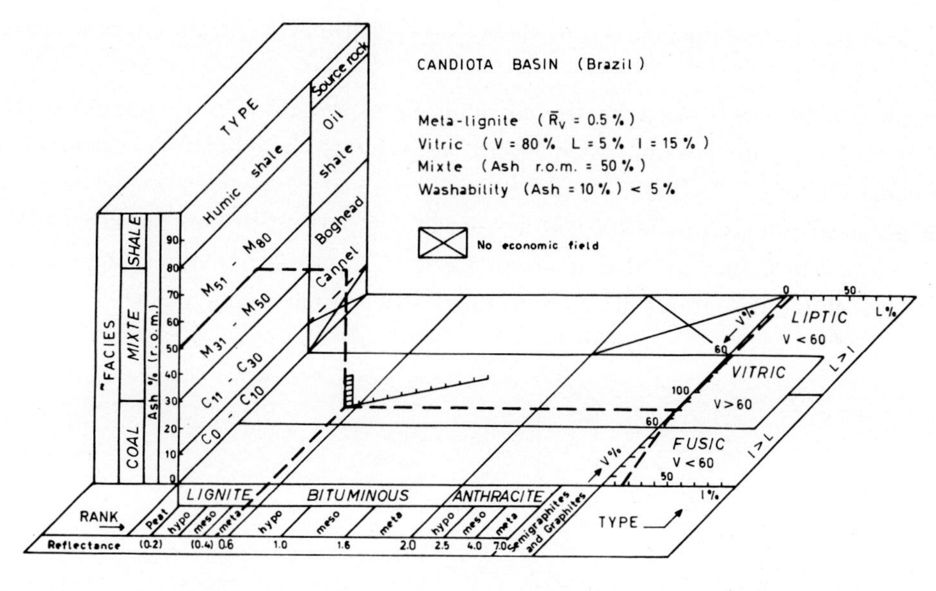

Fig. 5 Example of a coal from the Candiota Basin (Brazil) plotted in the Aplern's System, detailed chart

CONSLUSIONS

The classification proposed covers all solid fossil fuels with the exception of peats, solid bitumens and graphites.

The chart uses only simple words and three main parameters: RANK, TYPE and FACIES for an overall but fairly complete characterisation. The properties used for this classification are: mean random reflectance of vitrinite for Rank, maceral constitution for Type, and ash content and washability potential for Facies. Graphic representations are provided in order to facilitate comparisons between basins.

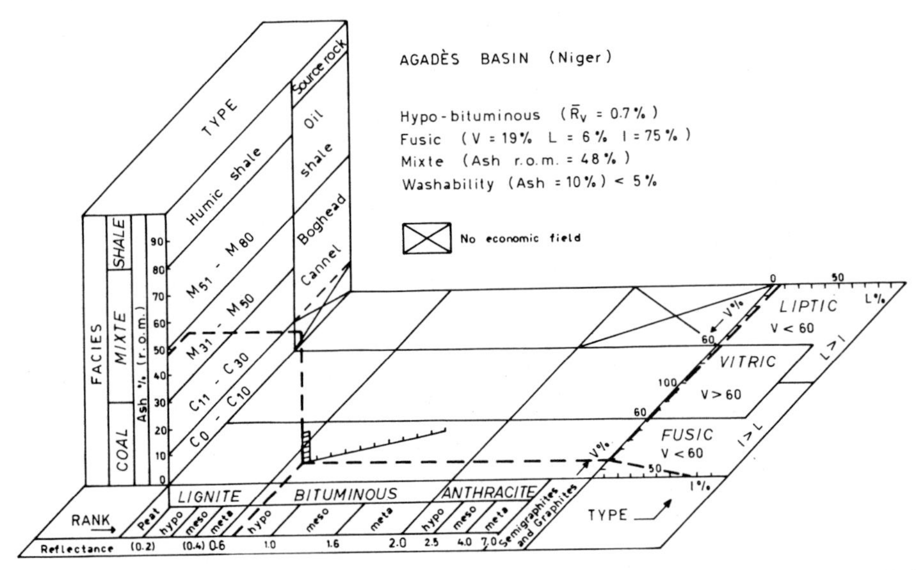

Fig. 6 Example of a coal from the Agadès Basin (Niger) plotted in the Alpern's System, detailed chart

A possibility is left open for the calculations of reserves on a common basis. Many other applications are possible with regard to geology, beneficiation, carbonization, *in situ* evaluation, etc.

A great advantage of the present system in comparison to the complex code number one developed in Geneva is that it permits a better mutual understanding, scientific exchanges, teaching and, above all, a correlation between national systems.

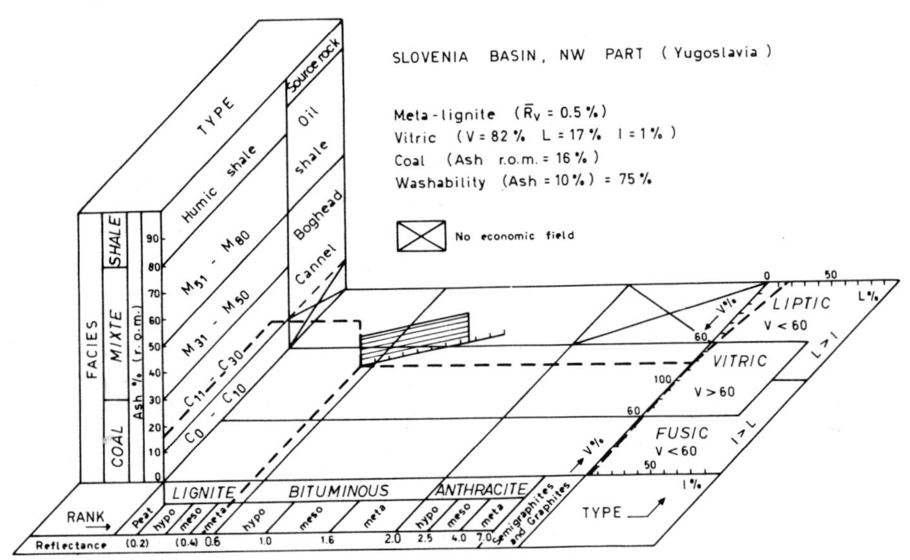

Fig. 7 Example of a coal from the Slovenis Basin (Yugoslavia) plotted in the Alpern's System, detailed chart

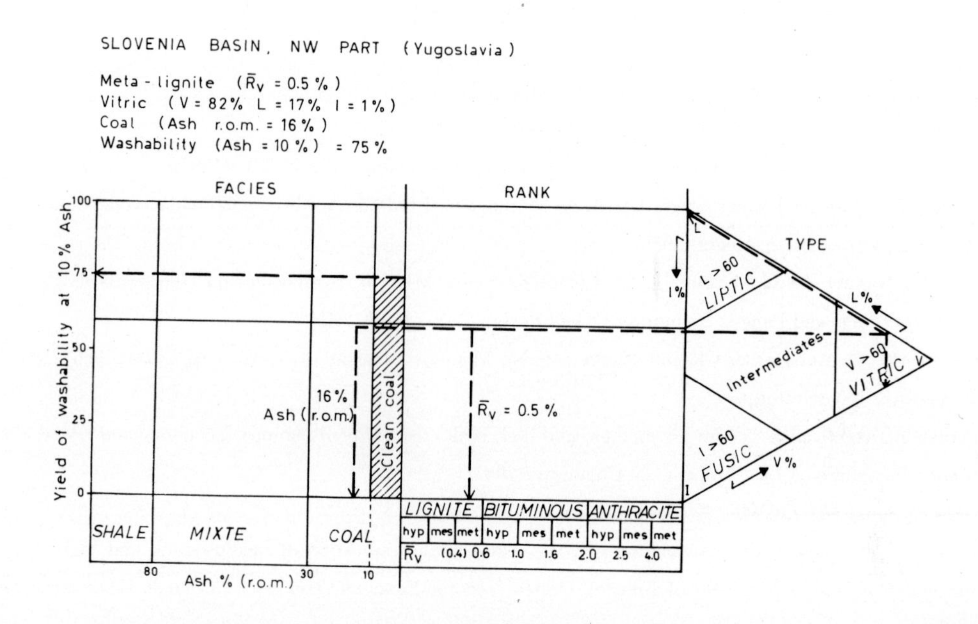

Fig. 8 Example of a coal from the Slovenia Basin (Yugoslavia) plotted in the Alpern's System, alternative chart

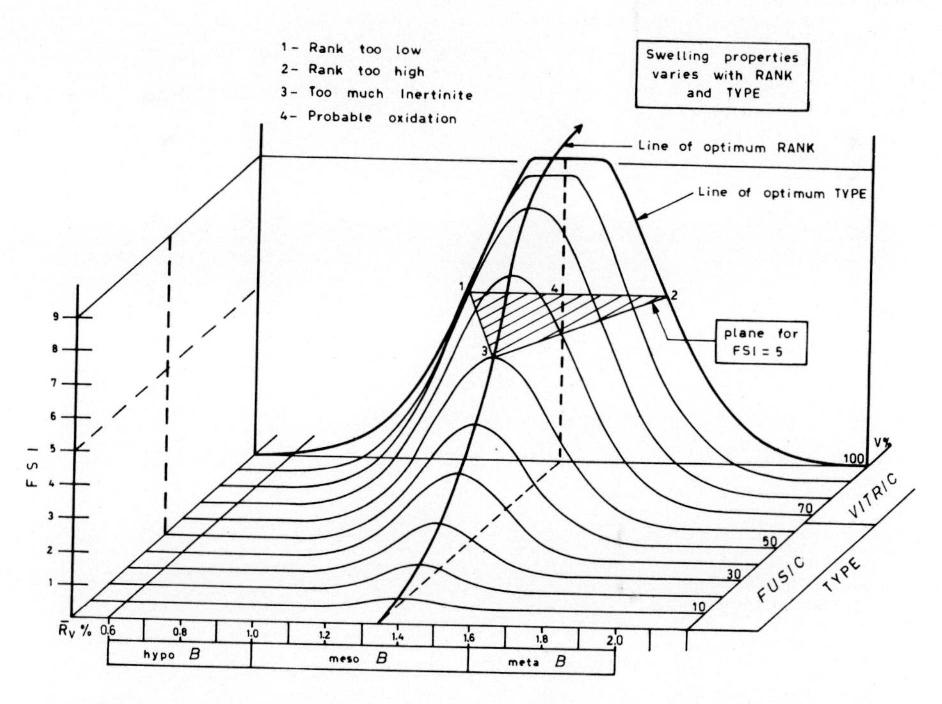

Fig. 9 Example of utilization of Alpern's System for definition of coking properties of clean coals

REFERENCES

1. Alpern B, Publ Tech Charbonn France, 3 (1979), 195.
2. ——, Bull Centres Rech Explor-Prod. Elf-Aquitaine, 5 (2) (1981), 271.
3. ——, Proc Sympos Advances in the Study of the Sydney Basin, Newcastle, N.S.W. (1987).
4. Alpern B and Nahuys J, Charbon du Bassin de Morungava, Etude de Six Forages. Fundacao de Ciencia e Tecnologia-CIENTEC, Porto Alegre (1985).
5. Alpern B, Nahuys J and Martinez L, In: Lemos de Sousa MI (ed), Symposium on Gondwana Coals, Lisbon, 1983. Proceedings and Papers. Comun Serv Geol Portg, 70 (2) (1984), 299.
6. Falcon RMS, In: Anhaeusser CR and Maske S (eds), Minerals Deposits of South S Africa, 2 (1986), 1899. Geol Soc S Africa. Johannesburg.
7. International codification system for medium and high rank coals. United Nations, Economic and Social Council, Economic Commission for Europe, Coal Committee(1987).
8. Lemos de Sousa MJ, Bol Min, 22 (3) (1985), 1.
9. Preparation of a new ECE classification of coals. Draft sub-classification of medium-rank and high-rank coals (transmitted by the Government of France). United Nations, Economic and Social Council, Economic Commission for Europe, Coal Committee (1986).
10. Standard Classification of Coals by Rank. ASTM Designation: D 388-82. In: Annual Book of ASTM Standards, Section 5—Petroleum Products, Lubricants, and Fossil Fuels, Vol. 05.05-Gaseous Fuels; Coal and Coke (1984), 242. ASTM, Philadelphia (Pa.).

XIe Congrès International de Stratigraphie et de Géologie du Carbonifère Beijing 1987,
Compte Rendu 1 (1991): 169-176

FOUNDERS OF THE PENNSYLVANIAN SYSTEM: 19TH CENTURY GEOLOGY IN THE U.S.A.

William M. Jordan
(Millersville University, PA 17551, USA)

ABSTRACT

Since 1953 the Carboniferous of Conybeare and Phillips (1822) has been recognized by the U.S. Geological Survey as consisting of two subdivisions of systemic rank, a Mississippian System and a Pennsylvanian System. However the origin of this North American usage and terminology dates from studies beginning in the first half of the 19th century.

In the upper Mississippi River valley, Owen (1839) recognized a lower carbonate sequence, which he named the "Subcarboniferous", that is separated from overlying coal-bearing clastic rocks by a major regional disconformity. These clastics were studied to the east by Henry D. Rogers and his assistants, and are described in the reports of the 1836-1842 First Geological Survey of Pennsylvania. In 1840 Rogers divided these coal measures into a lower conglomerate, a middle Allegheny series, and an upper Monogahela series. The Allegheny and Monongahela are named for the two tributaries which form the Ohio River at Pittsburgh, Pennsylvania. Modifications were later made by workers of the Second Geological Survey of Pennsylvania. Stevenson (1873) redefined the Monongahela. Platt (1875) separated the Conemaugh Group from Roger's Allegheny, and Lesley (1876), director of the survey, named the lower conglomerate Pottsville after a mining town in the anthracite district. In 1891 I.C. White named the topmost sequence above the redefined Monongahela the Dunkard Creek series. Thus the original sequence of conglomerate, lower productive coal measures, lower barren coal measures, upper productive coal measures, and upper barren coal measures had become in ascending order the Pottsville, Allegheny, Conemaugh, Monongahela, and Dunkard.

In 1870 the Subcarboniferous of Owen was renamed "Mississippian" by Alexander Winchell, then State Geologist of Michigan. Later, in 1891, in correlation papers published by the U.S. Geological Survey in order to regularize stratigraphic terminology, H. S. Williams designated the Mississippian a series equivalent to Lower Carboniferous, calling the Upper Carboniferous the "Pennsylvanian" series after the well known coal measures in Pennsylvania discussed above. Following publication of the 1906 Chamberlin and Salisbury *Textbook of Geology* which elevated the rank of Williams' Mississippian and Pennsylvanian series to systems, such designation became commonplace as the result of widespread American usage.

INTRODUCTION

The history of subdivision of the Carboniferous into Mississippian and Pennsylvanian Systems in North American practice and the delineation of various units within the original Pennsylvanian, along with the story of the various personalities involved, provides some interesting insights into the nature of American geology in the 19th century.

Application of the term "Carboniferous" to part of the stratigraphic succession in Great Britain by William Phillips and the Rev. W. D. Conybeare in 1822 has been described by William Ramsbottom[1] in the *Compte Rendue* of the 9th Congress, held in the United States. Since the 1906 publication in America of the very popular Chamberlin and Salisbury *Textbook of Geology,* the use of the terms Mississippian and Pennsylvania as two systems equivalent to the Carboniferous has been widespread in North America, although not officially adopted by the U.S. Geological Survey until 1953. The story of this transformation begins more than one hundred fifty years earlier in Indiana, on what was then the frontier of settlement in the northwest.

DAVID DALE OWEN AND THE INDIANA SURVEY

America in the early 19th century was expanding rapidly westward where new states were being formed beyond the boundaries of the old British colonies. In what was to shortly become Indiana, George Rapp, a German religious leader, in 1814 founded the town of Harmonie on the banks of the Wabash River, adjacent to what was to become the state of Illinois. In 1825 this town was sold to Robert Owen, the wealthy British industrialist and utopian social reformer who had established the model industrial community of New Lanark in Scotland.[2] Owen's town, renamed New Harmony, was intended to operate with a social order based on community ownership and equality of work and profit. His partner in this venture was William Maclure, known today as the "father of American geology"[3] because of the breadth of his geological activity and the preparation of the first extensive geologic map of the eastern U.S. Soon New Harmony, an outpost in the wilderness, became a center of naturalist and scientific activity that included geology, thanks to the influence of Maclure. While Owen's social experiment failed in 1827, New Harmony remained an educational, scientific, and cultural center until the disruptions of the American Civil War in mid-century.

In 1837, upon organization of the first geological survey of Indiana, David Dale Owen, son of Robert Owen, was appointed state geologist. The younger Owen, who was to become a distinguished geologist and who later conducted several mineral and land surveys in the upper Mississippi Valley for the U.S. Government,[4] was trained both in Europe and by means of field work with New Harmony geologists such as Gerard Troost. The Indiana Survey included preparation of maps and cross-sections (now missing) but was basically of a reconnaissance nature, involving extension westward of the known stratigraphy of the state of Ohio where the coal-bearing rocks had recently been described by Hildreth.[5]

At Oil Creek in Perry County, Indiana, adjacent to the Ohio River which is tributary to the Mississippi, Owen recognized a series of partially oolitic limestones immediately below coal-bearing rocks. These he named Subcarboniferous "as indicating its position immediately beneath the coal, or Carboniferous group of Indiana"[6] and in a footnote elaborated as follows: "The fossils generally coincide closely with those of the Carboniferous or Mountain limestone of Europe; but as no perfect seams of coal have ever been yet observed alternating with those deposits in this country, and as most of its fossils differ decidedly from those of the coal formation, it would seem to preclude the possibility of including it, here at least, as some European geologists do their Mountain limestone, in the Carboniferous group..., I prefer designating it by the term 'Subcarboniferous', which merely indicates its position beneath the Carboniferous group without involving any theory." The contact observed by Owen is now known to be a widespread regional disconformity. Owen was responsible for "recognition of the value of the fossil forms *Pentremites golbosa* and *Archimedes* as horizon markers, as fixing the age of the oolitic limestone older than the Coal measures, notwithstanding its lithological resemblance to the European Jurassic, and therefore marking the lower limit of the coal beds" which was an important stratigraphic generalization for the period.[7]

FIRST GEOLOGICAL SURVEY OF PENNSYLVANIA

The decade of the 1830s was, in fact, a time of establishment of geological surveys in many American states. This included New York where paleontologist and biostratigrapher James Hall came to fame, and Pennsylvania with its great resources of both anthracite and bituminous coal. In 1836 Henry D. Rogers, a native of Philadelphia who had been trained in geology in Europe along with David Owen, and was then professor of geology and mineralogy at the University of Pennsylvania as well as being state geologist of New Jersey, was appointed to head a geological survey of Pennsylvania.[8] It was a vast undertaking to analyze and record the geology of this large, geologically complex area of the central Appalachians, a good part of which was still forested wilderness. Rogers is best known today, however, along with his brother William B. Rogers, then state geologist of neighboring Virginia, for development of concepts in structural geology based on observations made in the course of their survey work. Henry Rogers eventually went on to become Professor of Natural History at the University of Glasgow, Scotland, and was the first American to obtain such a prestigious overseas academic appointment.

The field work of the first geological survey of Pennsylvania was conducted from 1836 to 1842, but its final report was not completed until 1847, nor published until 1858.[9] However six small annual reports were issued from 1836 through 1842. In 1840 Rogers[10] divided the coal-bearing rocks of Pennsylvania into a lower conglomerate, a middle Allegheny series, and an upper Monongahela series, the latter two subdivisions being named for the large tributaries that join to form the Ohio River at Pittsburgh. The base of this thick coal-bearing succession was defined not on fossils alone but "on the suddenness of the change from marine to terrestrial forms and the rapid appearance of that amazing vegetation characteristic of the coal period."[11]

The various coal basins of Pennsylvania are described in over 600 pages of Roger's profusely illustrated final report of 1858 which was, curiously for a government survey, published privately and issued in both Philadelphia and Great Britain. Forty-seven of its 1 631 pages are a description and discussion of the coal flora by Leo Lesquereux. In addition there are sections describing the coal fields of the U.S. and the British North American provinces, methods of prospecting for and mining coal as it was done in Pennsylvania, and the nature of foreign coal fields and the coal trade.

SECOND GEOLOGICAL SURVEY OF PENNSYLVANIA

In the middle of the 19th century, the American Civil War of 1860-1864 brought substantial changes to the economies of the northern states, as well as destruction to the south. In Pennsylvania a vast coal mining and iron and steel industry had grown up, along with a well developed transportation network.[12] This had first developed as canals were built utilizing water level routes facilitated by numerous water gaps through the Appalachian ridges, although an elaborate portage system was needed to cross the "Allegheny Front" which is the eastern edge of the Appalachian Plateau and the boundary of the bituminous coal fields. These canals were eventually supplanted by a vast network of railroads, especially the famed Pennsylvania Railroad which prided itself as being "the standard of the world". In addition, in 1859, the world's first drilled oil well was completed at Titusville in Pennsylvania.

This mineral-based prosperity resulted in calls for establishment of a Second Geological Survey in Pennsylvania and in 1874, J. Peter Lesley, an assistant on the First Survey under Rogers, was appointed state geologist. Lesley, an ordained minister, was then professor of geology at the University of Pennsylvania, as Rogers had been, and was similarly a native of Philadelphia. He had published a *Manual of Coal and its Topography*[13] in 1856, an *Iron Manufacturer's Guide*[14] in 1859 while serving as secretary of the American Iron Association, and was a founding member of the National Academy of Science.

Lesley's character and his role as a pioneer in understanding the detailed geology of the Pennsylvania anthracite basins has been described by me previously in the *Compte Rendue* of the 9th Congress.[15] Lesley was an extremely literate, articulate individual who was bitterly critical of what he felt was lack of recognition for his work, and that of other assistants, on the First Geological Survey under Rogers. His own survey operated on a large scale until 1887, itself employing numerous assistants. Among these were John J. Stevenson, Franklin Platt, and Israel C. White who, along with Lesley, were involved in partially redefining Rogers original subdivision of the coal measures.

One of these assistants, New York City born John J. Stevenson had been professor of chemistry and natural sciences at the University of West Virginia and an assistant on the Geological Survey of Ohio before joining the Second Pennsylvania Survey in 1875.

In a paper based on his Ohio work,[17] Stevenson redefined Roger's Monongahela series by giving it a specific upper limit, the top of the Waynesburg coal. The underlying lower barren coal measures beneath the Pittsburgh coal contains a widespread marine unit, the Ames Limestone, and had been part of Rogers's Allegheny series. It was redefined[18] in 1875 as the Conemaugh series by another survey assistant Franklin Platt. The Conemaugh River for which the series is named is a tributary of the Allegheny River, entering above Pittsburgh, and was utilized by the Pennsylvania canal system and by the Pennsylvania Railroad along part of its route. Platt, before joining the Pennsylvania Survey, had been a member of the U.S. Coast Survey and, during the American Civil War, was chief engineer of the military division of the Mississippi.

In 1876 J. P. Lesley, director of the Second Survey, named the lower conglomerate of Roger's sequence "Pottsville" after a prominent mining town of that name in the easternmost anthracite district. The final modification in the 19th century of Roger's original subdivision of the coal measures came in 1891. It was made by Israel C. White who had been a student of John Stevenson, an assistant on the Second Geological Survey of Pennsylvania, a U.S. Geological Survey geologist, and was at the time professor of geology at the University of West Virginia. He renamed the topmost sequence lying above the Monongahela (as redefined by Stevenson), which was the former "upper barren coal measures" of Rogers, the Dunkard Creek series.[20] Thus by the end of the century the original succession of conglomerate, lower productive coal measures, lower barren coal measures, upper productive coal measures, and upper barren coal measures had become, in ascending order, the Pottsville, Allegheny, Conemaugh, Monongahela, and Dunkard. I.C. White became director of the West Virginia Geological Survey in 1897 and held that post for the next thirty years and recently, although suggested as early as 1962,[21] an alternative stratotype for the Pennsylvanian has been proposed based on the continuous sequence that is present within the boundaries of West Virginia.[22]

ALEXANDER WINCHELL AND THE MISSISSIPPIAN

The story of the transformation of Owen's "Subcarboniferous" and Roger's coal measures sequence, as eventually modified, into Mississippian and Pennsylvanian Systems returns now to the midwestern states. In 1870, Alexander Winchell, then state geologist of Michigan and concurrently professor of geology, zoology, and botany at the University of Michigan, introduced the term Mississippi limestone series, or "Mississippi group", in a lengthy paper on the age and correlation of the Marshall group, a sequence of clastic rocks found in Michigan. In a footnote contained in this paper[23] Winchell states that "I propose the use of this term as a geographical designation for the Carboniferous limestones of the United States which are so largely developed in the valley of the Mississippi River."

While contributing the term Mississippian to geological usage, Winchell was best known however as a teacher, public lecturer, and popular science writer, as well as for his stand on organic evolution and for a "pre-Adamite" evolutionary history of mankind.[24] He taught briefly at Vanderbilt University in Tennessee, but had his position abolished from beneath him in 1878 because of these views which, for the time and region, were considered as being too radical. He subsequently returned to the University of Michigan and in 1888 Winchell, along with John Stevenson and others, was one of the those most directly responsible for the founding of the Geological Society of America.

HENRY S. WILLIAMS AND THE U.S.G.S. DEVONIAN AND CARBONIFEROUS CORRELATION PAPERS OF 1891

Another Geological Society of America founder was Henry S. Williams who enters the story in 1891. As an individual active in the organization of the geological profession in 19th century America, he was also involved in establishing both the Scientific Honor Society of Sigma Xi and the Paleontological Society. Williams was professor of geology and paleontology at Cornell University in New York, and an assistant geologist of the U.S. Geological Survey at the time when John Wesley Powell was director. One of the first tasks of the then new U.S. Survey was preparation of a set of "correlation papers" to regularize the usage of stratigraphic terminology. Williams was author of *Correlation Papers, Devonian and Carboniferous,* issued in 1891 as U.S.G.S. Bulletin 80.[25]

To justify expenditure of public funds on such a project, Powell had previously written[26] that, "in order to develop the geological history of the United States as a consistent whole, it is necessary to correlate the various local elements .... The correlation of strata separated by wide intervals of discontinuity can be effected only through the study of their contained fossils.... The study of the data and principles of correlation is thus seen to be a necessary part of the work of the Geological Survey, and by making the study at the present time it can offer a timely contribution to general geologic philosophy". In the letter of transmittal for Bulletin 80, written by Grove K. Gilbert who was the Geologist-in-Charge of the correlation project, Gilbert states that each correlation paper "should show into what series (major subdivisions) the system has been divided in various parts of North America" and "should show whether and to what extent the subdivisions of the system in any or all of its American provinces can be correlated with the subdivisions of the system in Europe."

Williams states in his 1891 *Correiation Papers, Devonian-Carboniferous,* referring to a portion of the calcareous rocks of the midwest, that they are characterized by Carboniferous fossils and indicates that they had been variously called "Mountain limestone", "Carboniferous limestone", "Subcarboniferous", and "Lower Carboniferous". He believed that no one of these names was entirely satisfactory, and further states that since "these formations are bound together by a common general fauna and constitute a conspicuous feature in the geology of this region, it is proposed to call them the Mississippian series." This series "is typically developed in the States forming the upper part of the valley of the Mississippi River, viz, Missouri, Illinois, and Iowa." Williams continues, saying that "the name is a slight modification in form and usage of a name proposed by Alexander Winchell in 1870.... I have already proposed the use of the name in this sense in recent reports to the State geologists of Arkansas and Missouri." Williams also states, in reference to earlier work on these rocks and their identification as Carboniferous, that "in the Mississippian province the identification of the rocks from the Coal Measures downward was correctly made, not because of accurate knowledge of the fossils, but because the three grand divisions of the typical English Carboniferous system were there present in the same order: first, a series of limestones, then conglomerate or sandstone, then Coal Measures."

With regard to the coal-bearing rocks, in the *Correlation Papers* Williams speaks of the "elaboration of the Pennsylvanian series of the Coal Measures" by the Second Geological Survey of Pennsylvania and he uses the terminology "The Coal Measures or Pennsylvanian Series" as a chapter title, as he does the phrase "The Lower Carboniferous or Mississippian Series." Thus in a manner similar to his use of Winchell's terminology of twenty-one years previously, the name "Pennsylvanian" was also transformed into a series of the American Carboniferous.

THE CHAMBERLIN-SALISBURY TEXTBOOK OF 1906

These terms, Mississippian and Pennsylvanian, remained as "series" in the usage of the United States Geological Survey until 1953. However in 1906 a three-volume textbook appeared, written by Thomas C. Chamberlin and Rollin D. Salisbury.[27] A one-volume condensation, called *A College Text-book of Geology,*[28] was issued in 1909 and was widely adopted by schools in the United States. Chamberlin, a student of Alexander Winchell, had been president of the University of Wisconsin and was then professor of geology at the University of Chicago. His prestige at the time was very great, in part because of his enunciation of a planetesimal hypothesis for the origin of the earth. Salisbury was head of the Department of Geography at the same institution. In their textbook under the heading "Reasons for regarding the Mississippian as a distinct system", they cited the regional unconformity present in the midwest between the Mississippian and Pennsylvanian sequences as justification for the elevation to system rank of the two series proposed by Williams. Chamberlin and Salisbury state in *A College Text-book of Geology* that "the wide-spread emergence, erosion, and subsequent submergence recorded by the unconformity... is just the sort of change which is held to separate periods, not epochs. Nowhere else in the whole course of the Paleozoic era are so great physical changes embraced within the limits of one period." Apparently most American geologists agreed and, beginning in 1906, the Mississippian and Pennsylvanian through widespread usage (except by the U.S.G.S.) enjoyed the status of independent systems that have replaced in North America the original Carboniferous of Conybeare and Phillips.

REFERENCES

1. Ramsbottom W, 9^{e} Congr Int Strat Geol Carb Washington and Champaign-Urbana, 1979, C R 1 (1984), 109.

2. Patton JB, In: Field Trips in Midwestern Geology, Indiana Geological Survey, 1 (1983), 225.
3. Merrill GP, The First One Hundred Years of American Geology, Yale Univ Press (1924).
4. Rabbitt MC, Minerals, Lands, and Geology for the Common Defence and General Welfare, U.S. Govt Printing Office, 1 (1979).
5. Hildreth SP, Amer Jour Sci, N S, 29 (1836), 1.
6. Owen DD, Report of a Geological Reconnaissance of the State of Indiana; Made in the Year 1837, Osborn and Willetts (1839).
7. Merrill GP, op cit, 195.
8. Millbrooke A, Northeastern Geology, 3 (1981), 71.
9. Rogers HD, The Geology of Pennsylvania, a Government Survey, lippincott (1858).
10. Rogers HD, Fourth Annual Report on the Geological Exploration of the State of Pennsylvania, Harrisburg (1840).
11. Merrill GP, op cit, 377.
12. Jordan WM, In: Two Hundred Years of Geology in America, Univ Press of New England (1979), 91.
13. Lesley JP, Manual of Coal and its Topography, Lippincott (1856).
14. Lesley JP, The Iron Manufacturer's Guide to the Furnaces, Forges and Rolling Mills in the United States, Wiley (1859).
15. Jordan WM, 9^{e} Congr Int Strat Geol Carb Washington and Champaign-Urbana, 1979, C R 1 (1984), 121.
16. Jordan WM and Pierce NA, Northeastern Geology, 3 (1981), 75.
17. Stevenson JJ, Amer Philos Soc Trans, N S, 15 (1873).
18. Platt F, Report of Progress in the Clearfield and Jefferson District of the Bituminous Coal Fields of Western Pennsylvania, 2nd Penna Geol Surv, Report H (1875).
19. Lesley JP, The Boyd's Hill Gas Well at Pittsburgh, 2nd Penna Geol Surv, Report L, Appendix E (1876).
20. White IC, Stratigraphy of the Bituminous Coal Field of Pennsylvania, Ohio, and West Virginia, US Geol Surv Bull 65 (1891).
21. Branson CC, In: Pennsylvanian System in the United States, a Symposium, Amer Assoc Petrol Geol (1962), 97.
22. Englund KJ et al, Proposed Pennsylvanian System Stratotype, Virginia and West Virginia, American Geological Institute (1979).
23. Winchell A, Amer Phil Soc Proc, 11 (1870), 79.
24. Merrill GK, op cit, 395.
25. Williams HS, Correlation Papers, Devonian and Carboniferous, US Geol Surv Bull 80 (1891).
26. Powell JW, Ninth Annual Report of the US Geological Survey, US Govt Printing Office (1888).
27. Chamberlin TC and Salisbury RD, Textbook of Geology, Holt (1906), 3 vols.
28. Chamberlin TC and Salisbury RD, A College Text-book of Geology, Holt (1909), 602.

XIe Congrès International de Stratigraphie et de Géologie du Carbonifère Beijing 1987,
Compte Rendu 1 (1991): 177-186

CARBONIFEROUS PALAEOGEOGRAPHIC DEVELOPMENT IN CENTRAL EUROPE

Eva Paproth
(Geologisches Landesamt Nordrhein-Westfalen, Krefeld, FRG)

The Carboniferous palaeogeographic development of central Europe was strongly influenced by its tectonic setting (Fig. 1):

The oldest element and therefore perhaps the most stable was the East-European Platform. The platform was locally deformed as for instance, in the Baltic syneklise, the Moscow depression and the Pripyat-Donbas graben. These structural features are reflected in the palaeogeography. The more the platform border is approached, the more it is broken, first by important lineaments, then into larger and smaller blocks which are also reflected in the palaeogeographical development.

In northern Europe, the Caledonian orogeny had affected the platform border in the Early Palaeozoic.

In the Late Palaeozoic, the Devonian and Carboniferous, the platform border was affected by the Variscan or Hercynian orogeny; the axis of this orogene is marked on Fig. 1.

In the Late Mesozoic, the Alpidic orogeny attacked the platform border and affected the southern flank of the Variscan tectogene. Observations on the development of the Variscan tectogene are therefore concentrated on its northern flank.

The Variscan orogeny started at the beginning of the Devonian or somewhat earlier, and ended at about the end of the Carboniferous. For palaeogeographic reconstructions, the most important feature in this development is the contemporaneity of sedimentation and deformation. —No sedimentation is possible without deformation; this is trivial, but worth mentioning. A depositional area must arrive in a lower position than the erosional area; and to continue sedimentation, it is necessary to continue deformation. It is therefore impossible to distinguish between a depositional history of a tectogene and its deformation history.

Ten Carboniferous palaeogeographical maps demonstrate the growth of the Variscan tectogene in central Europe which determined the Devonian and Carboniferous palaeogeographic development (Figs. 2 to 11).*

During the Carboniferous, the Variscan tectogene suffered its more dramatic phases. The development started, however, with the Devonian (or earlier).

---

* Figs. 2 to 11 after references 6,7 and several other sources.

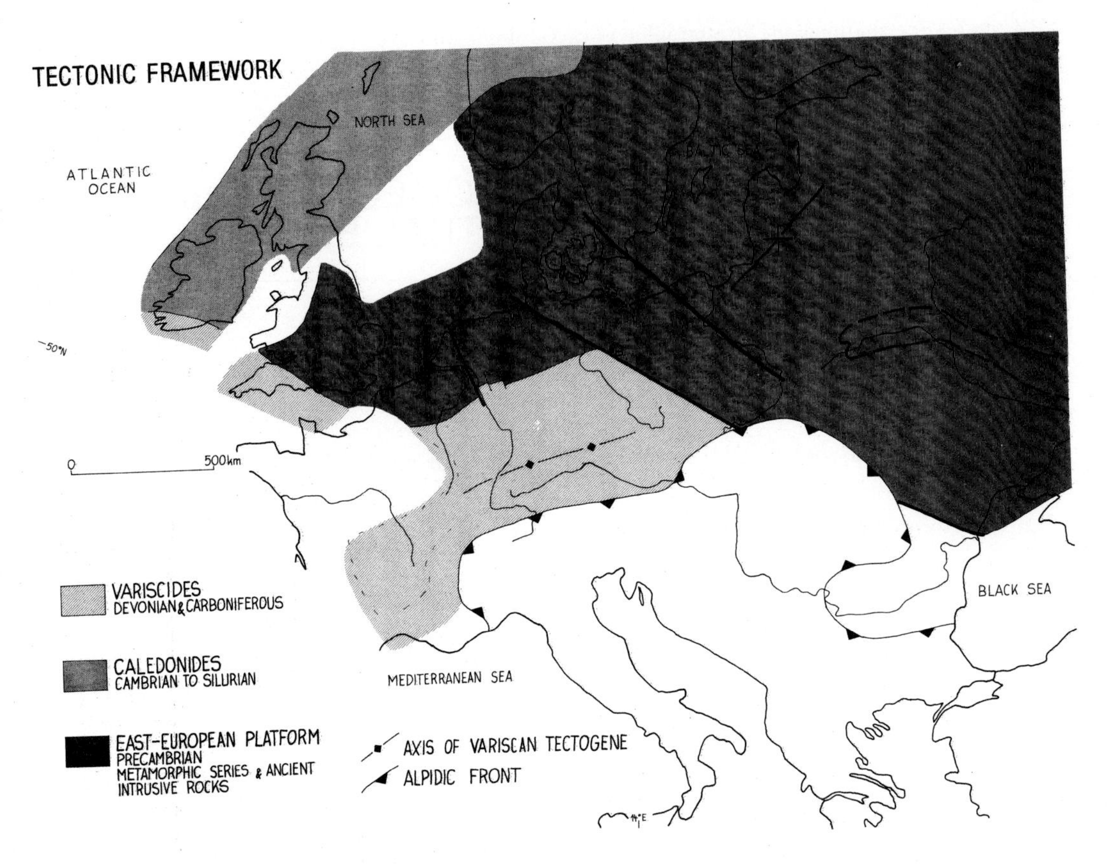

Fig. 1 Tectonic setting of central Europe

From west to east are marked: the NW/SE directed Mid Netherlands-Krefeld lineament; the NW/SE directed Hamburg-Dobruja lineament, B.=Town of Berlin; the NW/SE directed Tornguist lineament, running from Scania to the Wisla River; the Baltic syneklise; the Pripyat-Donez graben; M.=Town of Moscow in the Moscow depression in the north: O.=Town of Oslo.

In the beginning, in the area considered here, there was a huge "Old Red Continent", formed by the East European platform and by the Caledonides. A shallow sea covered the platform in the southeast and approached central Europe from there. This was facilitated by the geographic position of central Europe in the Devonian and Carboniferous:[1] it lay about 30^0 south of the equator in the Middle Devonian, travelled northward and reached the equator in the Late Carboniferous when the luxuriant plant production lead to the formation of numerous and thick coal seams. So the transgressional tendency from the southeast was facilitated by the south equatorial current that passed the area from the southeast to the northwest.

In the Devonian, the main sea current went through the area of the later Variscan Internides. But the Variscides grew, from the axial internal part to the external parts. In the Late Devonian already, the area of the Variscan Internides became more or less impassable because too many continental areas, islands and shoals had grown. The sea current from the southeast — or what was left of it — had to use a deviation via southern parts of the Old Red Continent.

At the beginning of the Carboniferous (Fig. 2) this deviation became a broad and shallow sea way; it was met by a shallow marine transgression from the west. The open sea lay in the south: the shelf of the Tethys covered an area which is now observable from the southern Alps to the Black Sea (and further east and west, of course, outside the maps).

Fig. 2 Tournaisian palaeogeography

Central Europe was already then destined to become dry, because the growing Variscides isolated this broad but shallow sea way more and more.

In the shallow sea, areas of restricted facies were widespread: in the carbonate dominated consolidated belt (on the East European Platform and the Caledonides), the Carboniferous limestone facies, there are evaporites — on the lee side of morphological highs. In the noncarbonate belt, the Kulm facies, thin black shales are common.

These thin black shales may be "diluted" by imported clastics, e.g. greywacke material. The flyschoid greywackes mark the acme of the tectogene, when the uplift of the hinterland reached a maximum, and unweathered, freshly eroded detritus was deposited in rapidly subsiding basins.

With the progressive uplift of the Variscides, proximal parts of the detritus fans were incorporated in the fold belt and the detritus fans shifted to the north (compare the position of the town Berlin, "B." in Figs. 2 to 11, and the shifting greywacke belt).

"Paralic" means "near to the sea" and represents in Figs. 2 to 10 all environments from brackish water to continental with marine ingressions. — In the Tournaisian, there was not yet a "limnic environment with coal formation".

The early Visean (Fig. 3) was marked by a regression on the Russian Platform. Linked to this were possibly the coal formation conditions in the Moscow Basin. — The main connection to the open Tethys sea was from the southeast.

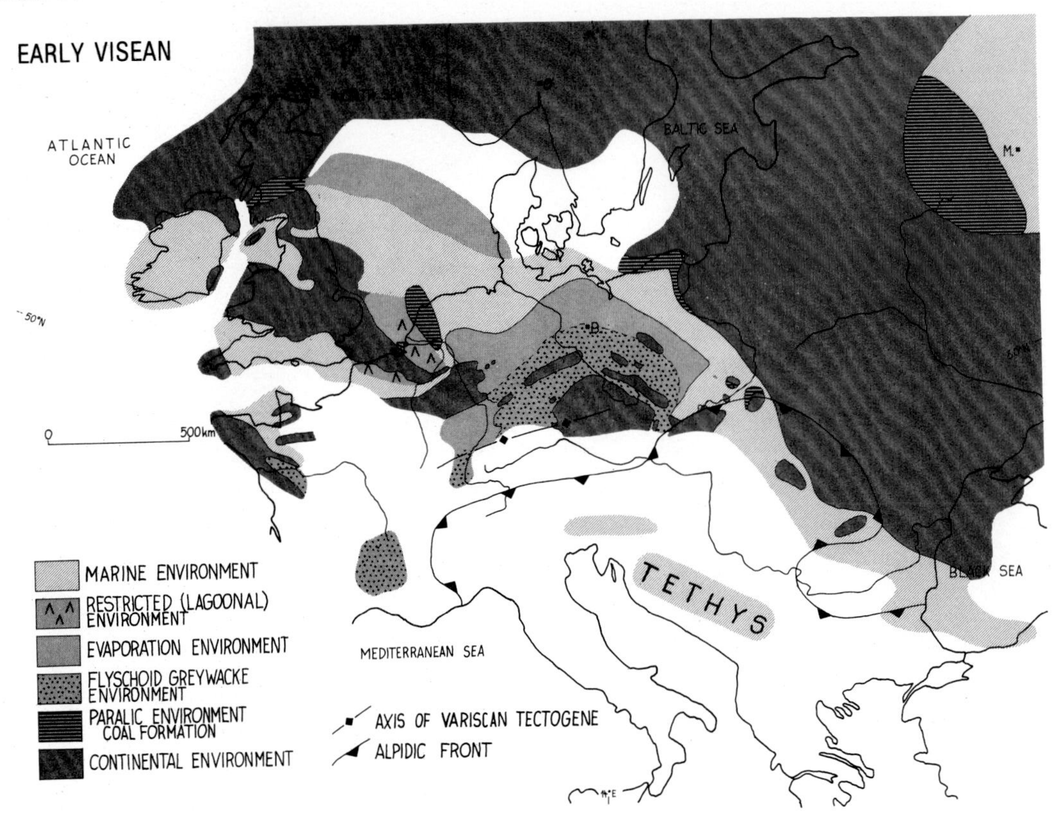

Fig. 3 Early Visean palaeogeography

In the middle Visean (Fig. 4) the progressive uplift of the Variscides is evident, it was followed by a further northward shift of the greywacke fan. In the Sudetes area (about 250km southeast of "B." = Berlin), a paralic environment with coal formation indicates that subsidence and detrital influx became equilibrated. In the east, the sea covered again vast parts of the platform. The Visean coal seams continued in the Moscow Basin and started in the Lublin area (the NW/SE running belt that lined the continental area of the Russian Platform). In the west, an easier passage of shallow marine waters crossed Britain.

The middle Visean tendencies continued into the late Visean (Fig. 5). This was the last test time at which there was a marine passage from the East European Moscow and Pripyat-Donez Basins to the central European area. Paralic environment became wide-

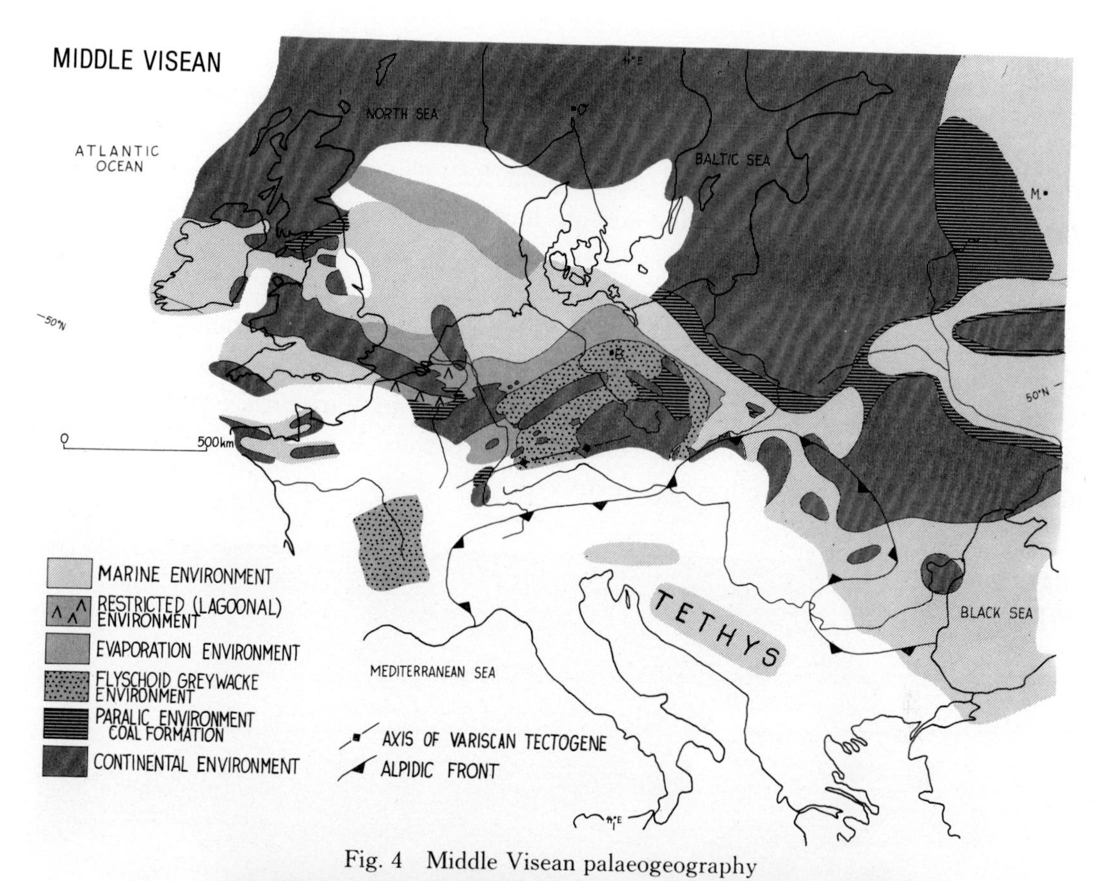

Fig. 4 Middle Visean palaeogeography

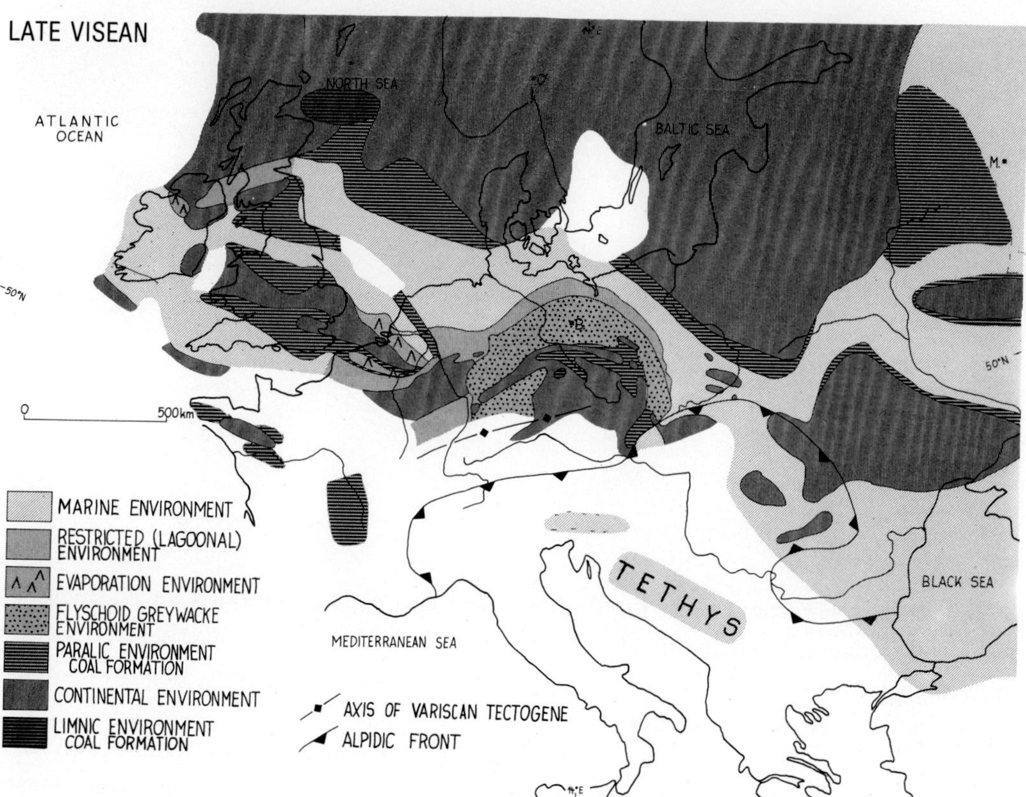

Fig. 5 Late Visean palaeogeography

spread, particularly in the North Sea area of northwestern Europe; a possibly limnic basin developed there. — Evaporite formation occurred for the last time in this Variscan framework, in Ireland and Belgium (again in the lee of highs). The next younger evaporites in this area are of Permian age and are linked to a fundamentally different tectonic and palaeogeographic situation.

The first limnic coal basin inside the Variscides (Karl-Marx-Stadt) is an important marker: in the top of the folded and uplifted Variscan Internides, "intramontane" graben structures gathered coarse siliciclastics and plant material for thick coal seams.

From the Namurian on (Fig. 6), the central European Variscides formed a complex body. The old, long lasting marine connection to eastern Europe disappeared, and the marine connection to the Tethys was restricted.

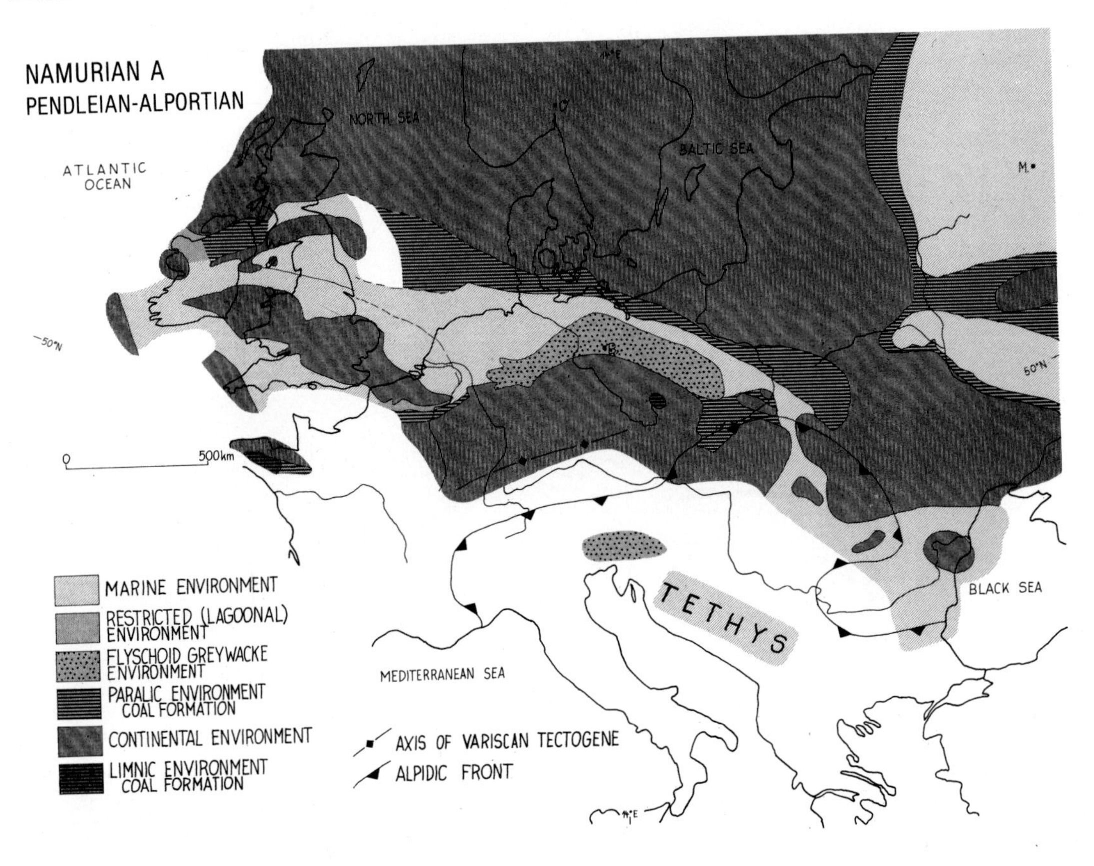

Fig. 6 Namurian A (Pendleian to Alportian) palaeogeography

Before the Namurian, most rapid subsidence and sedimentation (that is, thickest deposits) were concentrated in the flysch belt. From the Namurian on, thickest sediments were connected with coal formation (Upper Silesia). Coal formation required a balance between subsidence and deposition.

In the middle and late Namurian (Fig. 7), the south-easterly connection to the Tethys sea was finally cut off. The marine incursions came then from some westerly direction. They produced marine horizons of only few metres thickness, but wide distribution (up to about 2 000km). This required a uniformity of the paralic area in the "northwest European coal basin" that is difficult to imagine. —The Upper Silesian coal basin proved its unique character by continuing coal formation in limnic conditions.

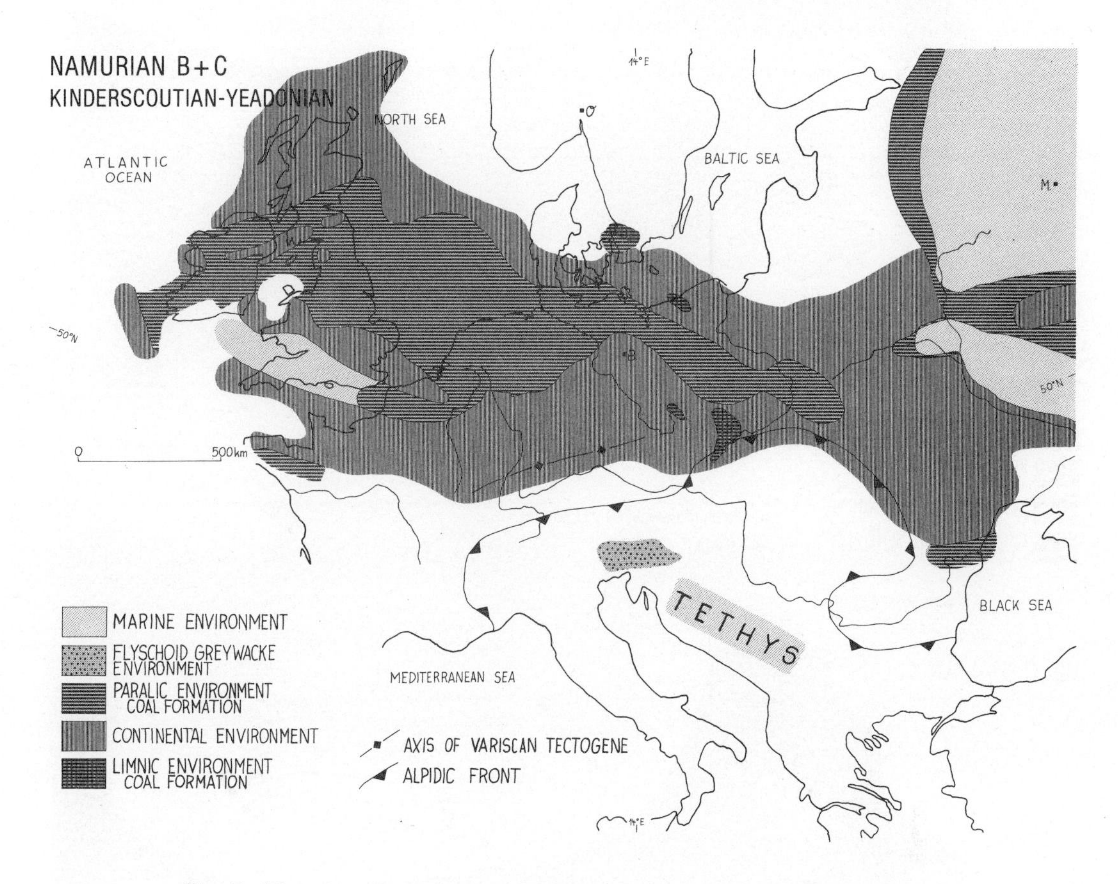

Fig. 7 Namurian B/C (Kinderscoutian to Yeadonian) palaeogeography

The Westphalian A/B (Fig. 8) was the time of maximum coal formation in the paralic area of northwestern Europe. During this time also, the number and importance of limnic intramontane, graben-like basins grew.

Still enigmatic is the occurrence of thin, marine, foraminifer-bearing carbonates in the Oslo area, as erosional remnants of Westphalian age (late Bashkirian and Moscovian fusulinid carbonates).[2] Since they were deposited when the marine influence decreased to nearly zero in central Europe (Fig. 9), it is improbable that the Oslo area had any connection to the sea via the paralic coal basin. A Late Carboniferous sea-way, connecting the Olso area with the Moscow Basin, appears possible. A pathway through southern Sweden, the Baltic syneklise, the Finnish Bay or the Gulf of Bothnia from the Arctic, where appropriate facies were distributed, may be plausible.[3]

From the Westphalian C on, sediments are thicker inside the Variscides, in large graben-like intramontane basins, than in the paralic belt.[4]

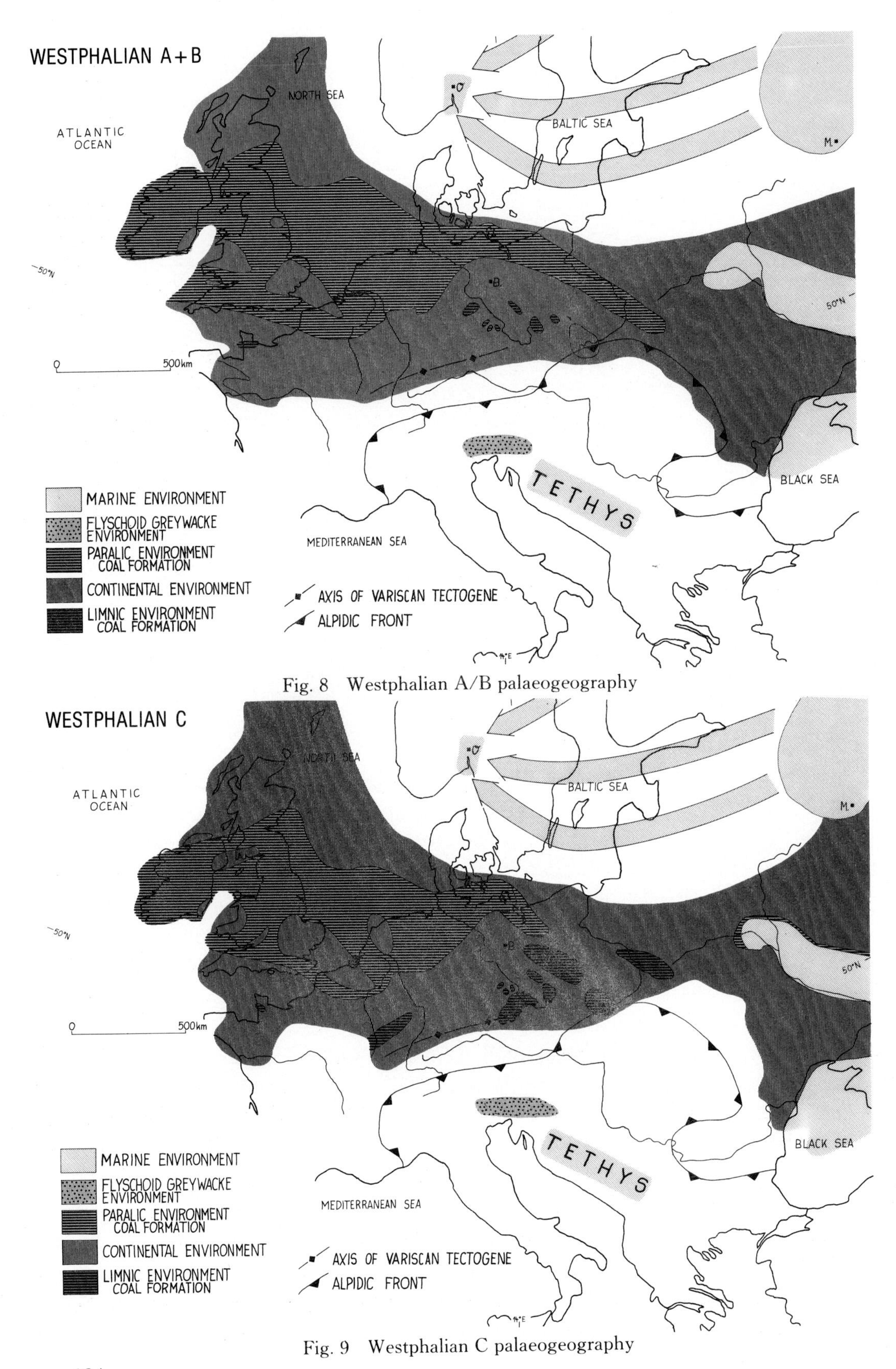

Fig. 8 Westphalian A/B palaeogeography

Fig. 9 Westphalian C palaeogeography

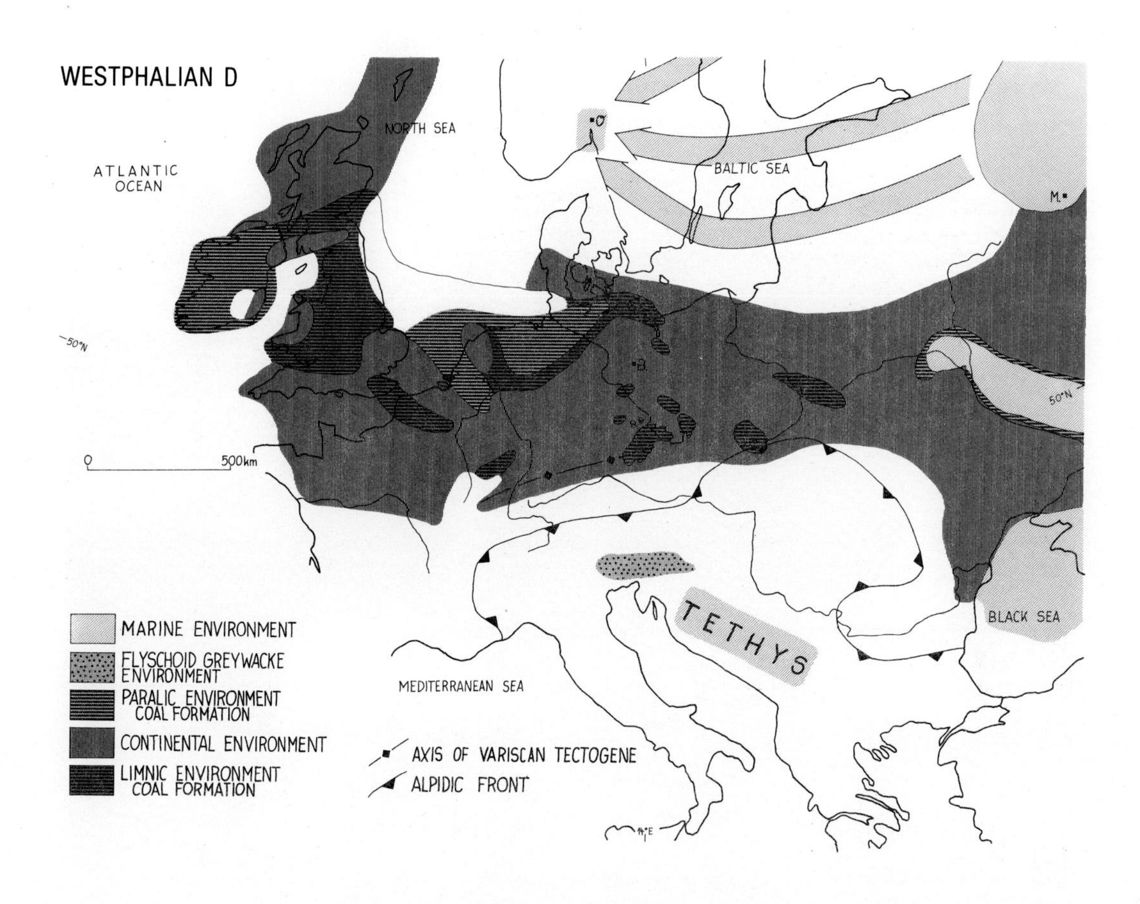

Fig. 10 Westphalian D palaeogeography

In the Westphalian D (Fig. 10), the retreating tendency of the northwestern European paralic basin is obvious. Siliciclastics are thin on the forelands, which implies that subsidence nearly stopped. This is in contrast to great thicknesses in the intra-Variscan graben-like limnic basins.

The Variscan development was finished in the youngest Carboniferous, the Stephanian (Fig. 11). Subsidence and deposition continued in numerous intramontane basins. The end of the Variscan development may also be seen by the lack of flyschoid greywacke and even paralic environment. —The beginning of the new Mesozoic cycle was marked by a volcanic paroxysm, the "Rotliegend-Vulkanismus".[5]

A general feature that emanates from the maps (Figs. 1 to 11)is the correlation of coal formation with a consolidated basement. In the whole of central Europe, coal seams formed either on the East European Platform and the Caledonian area, or in the intramontane basins which developed as grabens in the previously folded and consolidated Variscides. This certainly did not restrict later deformations.

Another general feature that emanates from the maps is the usefulness of international scientific cooperation: most of the maps and their essential parts have been sketched in the I.U.G.S. Project 86 "East European Platform (S.W. Border)",[6] and it is a most agreeable duty to acknowledge this and to thank very much the learned cooperators of this Project. I particularly want to thank those friends who dedicated time, effort and their invaluable knowledge into these maps.

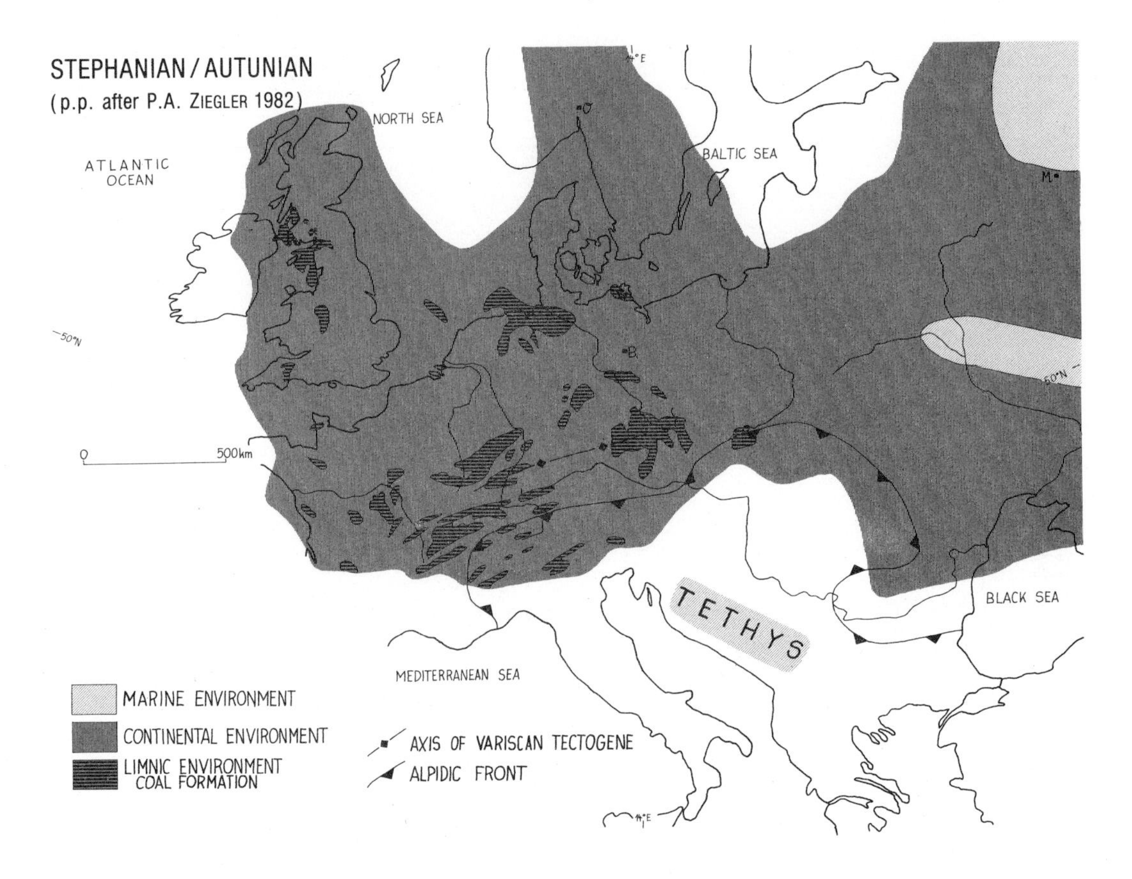

Fig. 11 Stephanian/Autunian palaeogeography (pro part after reference 8)

REFERENCES

1. Bless MJM, Bouckaert J and Paproth E, Bull Soc belge Geol, 93 (1-2) (1984), 189.
2. Olaussen S, Geol Mag, 118 (3) (1985), 281.
3. Bergström J, Bless MJM and Paproth E, Z dt geol Ges, 136 (1985), 181.
4. Bless MJM, Bouckaert J, Calver MA, Graulich JM and Paproth E, Mededel Rijks geol Dienst, 28 (1977), 101.
5. Plein E, Z dt geol Ges, 129 (1978), 71.
6. Jubitz K-B (ed), International Geological Correlation Programme Project 86 "East European Platform (S.W. Border)", Lithological-Palaeogeographical Maps Upper Visean, Westphalian A/B (in press).
7. Vinogradov AP (ed), Atlas of the Lithological-Paleogeographical Maps of the USSR, vol. 2, Vsesujusnyi aerogeologitsheskii trest, Ministerstva geologii SSSR (1969).
8. Ziegler PA, Geological Atlas of Western and Central Europe, Shell International Petroleum Maatschappij B.V. (1982).

XIe Congrès International de Stratigraphie et de Géologie du Carbonifère Beijing 1987, Compte Rendu 1 (1991): 187-200

FROM BIOSTRATIGRAPHY TO BIOCHRONOLOGY: TIME CORRELATION BY FOSSILS

Jürgen Remane
(University of Neuchâtel, Institut de Géologie, Emile-Argand 11, Ch-2000 Neuchâtel 7, Switzerland)

INTRODUCTION

1. General problems and definitions

1) Implications of the GSSP concept

As you may have seen from the preceding paper, considerable progress has been made during the last years concerning the definition of chronostratigraphic boundaries by reference points in a rock sequence, named GSSP.[1]

Before coming to my own subject, I would like to stress two implications of this procedure which seem important to me:

(1) According to the GSSP concept, chronostratigraphic units are not defined through their contents but by their lower boundaries. If we happen to discover rocks of a hitherto unknown time interval, there can thus be no doubt about their age in terms of the existing geological time scale. This is obviously a great advantage.

(2) Once a boundary is defined by a GSSP, it is no longer tied to a specific marker event. All available methods may be used to extend laterally the chronohorizon fixed by a GSSP.

I have stressed these two points in order to draw your attention to an important methodological problem: the classical biostratigraphic approach in chronostratigraphy is quite different. Zones, mostly defined by their contents, are the basic unit of time correlation. Therefore the question arises if or how far this classical approach can be reconciled with the correlation technique resulting from the GSSP concept.

I hope to be able to prove that there is no major difficulty in this. The main problem is that important terms are used with different meanings by different workers and therefore there are apparent contradictions. So, to start with a clarification of the most important terms and basic concepts is necessary.

2) Definition of basic chronostratigraphic terms

One of the central concepts in the present context is the one of chronostratigraphic units. As defined by Hedberg[2] a chronostratigraphic unit is a body of rocks bounded by isochronous surfaces, it corresponds to the sediments laid down during a given time interval. Isochronous surfaces are time planes or chronohorizons. All points (in the rock) on such a surface are — by definition — of exactly the same age. If we try to follow isochrones in the field, this is time correlation or chronocorrelation.

There are many methods of chronocorrelation (which I do not want to enumerate here). The point is that time planes are not directly accessible to observation, they have to be inferred from field data supposed to be chronologically significant. All of our correlation methods give us only more or less good approximations of isochronous surfaces. In this sense chronostratigraphic units, despite the fact that they were conceived as quite material rock bodies, are purely theoretic (i.e. conceptual) units, they can only be approached but never be attained.

2. Different meanings of the term biostratigraphy

1) The original meaning of biostratigraphy

In classical stratigraphy, up to the first half of our century, the term biostratigraphy was used in the sense of chronostratigraphy — the latter term did not exist at that time. This may seem illogical, but can be explained by the fact that fossils provide — still nowadays — the best means of long range chronocorrelation—due to organic evolution. The basic unit of this "chronostratigraphic biostratigraphy" is the zone. Unfortunately the classical zone concept was never made absolutely clear from the theoretical point of view. Zones were considered as time units—corresponding in the most straightforward case to the supposed life span of its index species : but normally the total life span of a fossil species will not be documented in one section.

Many stratigraphers would not hesitate to attribute rocks to a given zone even in the absence of its index species—if there were other arguments in favour of such an age attribution.

A didactic scheme (Fig. 1) illustrates very well the philosophical background of the classical biostratigraphic approach. We are indeed often able to reconstitute empirically the succession of species in time although none of the available sections gives a complete record of this succession.

Another schematic example shows how successions of strata can be dated once the succession of species is known (Fig. 2). In a first sequence (B) we know that species 2 should occur between species 1 and species 3, but we do not know whether its absence at this point is due to ecological factors or to a gap.

Other outcrops (A and C) do contain species 2. Section A makes it improbable that sediments of species 2 age should miss in B. But C gives us the proof : the sandstone level is present in all 3 sections, and it is indeed of species 2 age. So we may conclude that the beds from above the bioclastic limestone in B up to the sandstone (included) belong to the species 2 zone, although the marker species is absent in this outcrop.

This is, of course, a highly simplified version of the classical biostratigraphic procedure. The aim was to show that classical biostratigraphy contains an important part of abstraction and extrapolation. The underlying premise is that due to organic evolution, each fossil species is characteristic of a specific interval of time.

Classical biostratigraphic zones (in the meaning as explained above) were not always made to coincide with the life span of a marker species, in other words : not all of them were total range zones. Very often we deal indeed with partial range zones, characterized by the combination of two or more species, their co-occurrence or successive appearance. This has led to the distinction of a great number of different "types" of biostratigraphic zones most of which are, however, redundant. I shall come back to this problem later on. The point which is important for the moment is that these zones, too, were meant to be time units, they were so to say "biochronostratigraphic" zones.

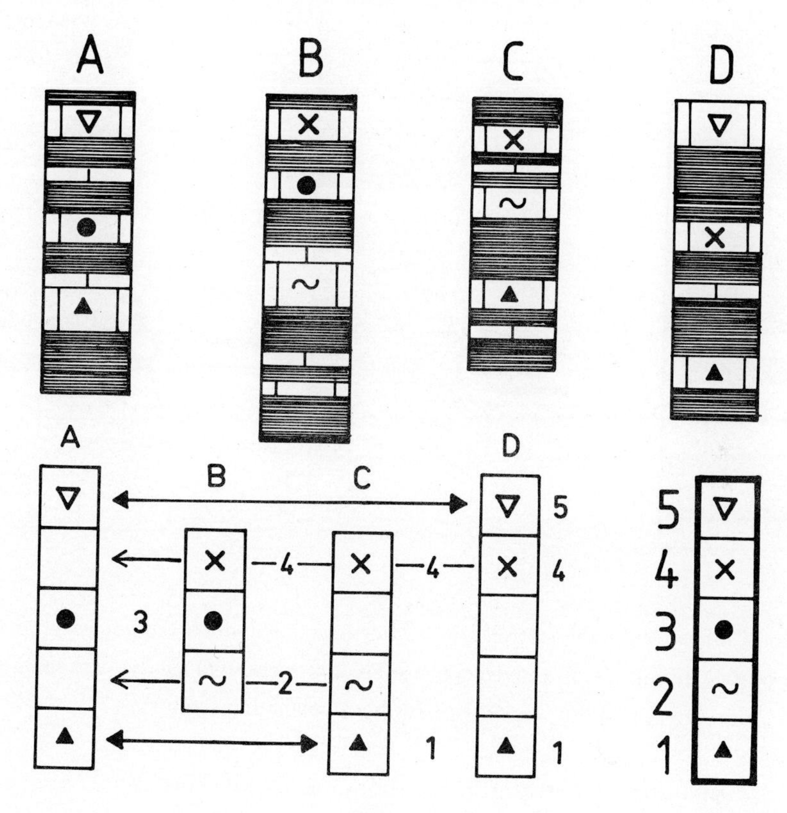

Fig. 1

Simplified scheme showing how a succession of species can be worked out from incomplete local biostratigraphic data. Above, 4 theoretic lithology with local successions of species, symbolized by geometric signs. Only 3 of the 5 species occur in one section, and it has been assumed that sedimentation rates vary from one section to another so that lithostratigraphic correlations are impossible. If one compares the local successions of species as in the schemes below, it appears that they complete each other mutually thus leading to the chronological scheme at the lower right.

2) Biostratigraphy according to the *ISG* (*International Stratigraphic Guide*)

(1) The problem with classical biostatigraphy is that the boundary between observation and interpretation is not always clear. There is often a tendency towards premature generalization. In such a case a zonation will only reflect some local or regional succession of species due to ecological factors, but not evolution.

In recent times it was mainly Hedberg[2] who postulated that biostratigraphic units should be reduced to their material base.[2] The zone of a species at a given place would then correspond to its known stratigraphic range, from its first up to its last occurrence.

With respect to their dimension in space, biostratigraphic zones in this strict sense are rock bodies defined by the presence of a characteristic species or association.

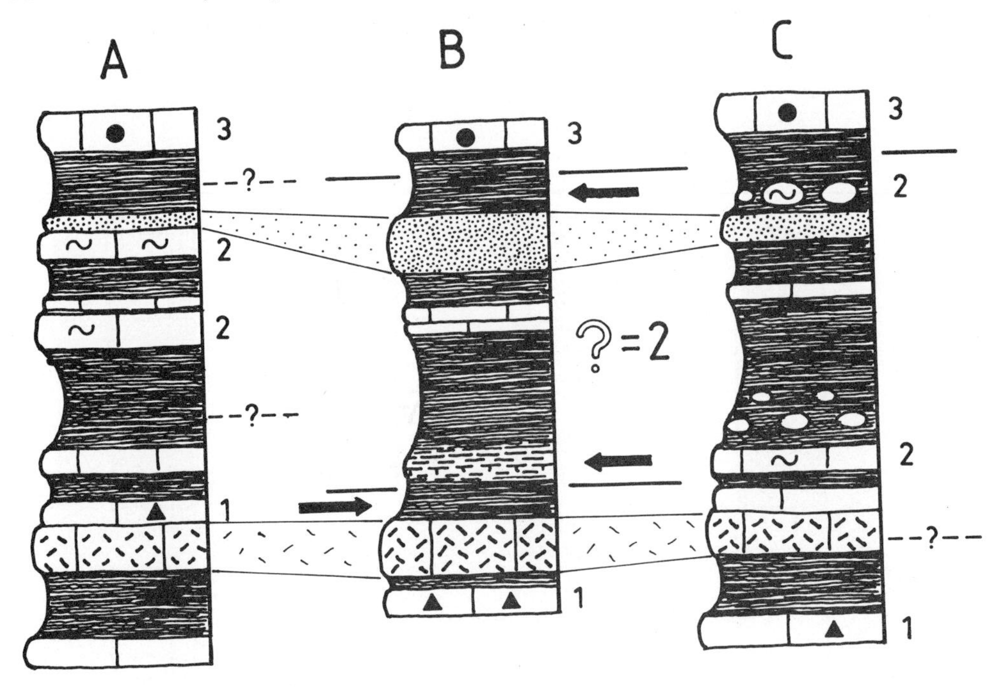

Fig. 2

With the help of lithostratigraphic correlation between sections which are not very far from each other, the presence of the chronozone of a given species (sp. 2 in section B) can be proven even in the absence of the marker (for further explanation see text).

Obviously material biostratigraphic boundaries of this kind are not necessarily isochronous (Fig. 3). As a matter of fact, the presence of a fossil species is limited by several factors which are independent from evolution. Or, to put it in an other way : a species may be really or apparently missing in sediments formed during its lifetime for several reasons :

(a) It did not live at that place for ecological or biogeographical reasons.

(b) The species was present but not fossilized or its remains were destroyed by diagenesis or metamorphism.

(c) There are fossil remains but they have not been found due to insufficient sampling.

This means in practice that the presence of a fossil species proves that the surrounding sediment was formed during the life time of this species — unless we deal with reworked material. But the opposite is not true : the absence of the marker may very well not be chronologically significant.

(2) It was certainly a good idea to use the term biostratigraphy in a strictly descriptive sense. But unfortunately all the different types of zones were incorporated in this purely descriptive biostratigraphy. As a matter of fact, the range zone of an individual

species (also called acrozone) is the only kind of zone which makes some sense in this context. Even here, formal zones are of limited value, because new discoveries will alter the known range of the species. In other words : the zone concept cannot be separated from its original chronologic meaning.

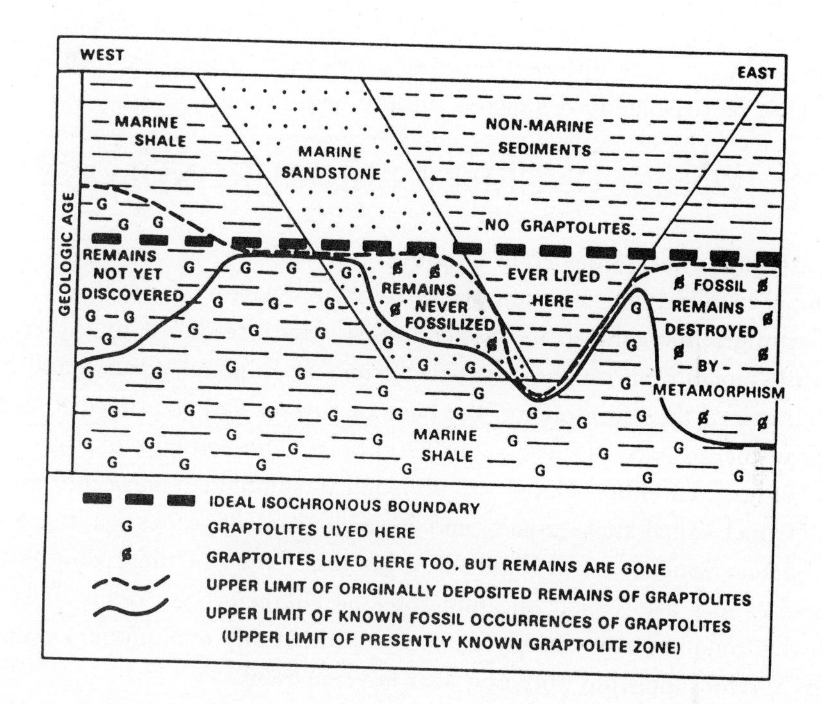

Fig. 3

Scheme showing all the factors which may cause diachrony of a biostratigraphic boundary.[2] Diachrony seems indeed to be very important, but if we take into account that the life time of a graptolite species is often no more than 1Ma, it appears that the simple presence of the species in two distant localities provides a much more accurate correlation than radiometric dating or any other method of chronocorrelation.

3) Biostratigraphy vs. biochronology

(1) I believe that there is a solution for this dilemma: the term biostratigraphy should indeed only be used in a purely descriptive sense. But the description of biostratigraphic field data needs no formal units like zones. All the observations can be perfectly presented by range charts and lithological columns or lithostratigraphic cross sections where biostratigraphic boundaries are indicated.

Unlike all other stratigraphic data, biostratigraphic data have real chronologic significance. They reflect, although in an indirect manner, organic evolution which is irreversible. Specific types of sediments may appear at any time when the neccessary physical conditions are realized, but a fossil species, once extinct will never appear again.

(2) Although we have abandoned the original chronostratigraphic meaning of the

term biostratigraphy, the underlying scientific method remains valid. But in order to separate clearly observation and interpretation we better speak now of biochronology. Biochronology corresponds to the chronologic interpretation of biostratigraphic raw data, and in this sense the term biochronology is already in current use.

So we finally arrive at the main subject of this paper: how can biochronology be organized, i.e. expressed in terms of formal units? Are zones in the classic sense still a valid concept? How many different types of zones may or must be distinguished? Or can biochronology be reasonably organized without using zones?

ZONE CONCEPT IN BIOCHRONOLOGY

1. Historical origin of the zone concept

1) Oppel's (1856-1858) zonation of the Jurassic System

The stratigraphic zone in its modern sense was introduced by Albert Oppel (1856-1858)[3] in his monograph on the Jurassic System of England, France and SW Germany. The correlation of these regions on the base of current stages proved to be very difficult for Oppel in the absence of clear stage definitions. The contents of a given stage varied from one author to another. Oppel was thus led to subdivide stages into their constitutive elements. Oppel called these zones, and he arrived at 33 zones for the 8 stages of the Jurassic System which were recognized at his time. Most of these zones were named after ammonites and also based on ammonites as characteristic fossils. It is interesting to note that this zonation was entirely empirical without any evolutionary premise. Darwin's *Origin of Species* appeared only one year later, in 1859.

According to Harland et al.[4] the Jurassic Period reaches from 144 to 213Ma BP, with a duration of 69Ma. This means that the mean duration of Oppel's zones is slightly more than 2Ma, which corresponds to a resolution of 1% for the Early Jurassic in 1858. Modern radiometric datings attain under favourable circumstances a resolution of 5%.

If we apply these values to the Carboniferous, the superiority of biochronology compared to radiometric dating becomes still more spectacular. A mean zone of 2Ma duration would then correspond to 0.7%, whereas the confidence interval of radiometric dating with 5% would correspond to 15Ma.

2) Difficulties with the Oppelian zone concept

(1) Oppel never gave a generalized definition of his zone concept. If we translate his zonal diagnoses into range charts of contiguous zones (as this has currently been done), we obtain the following image (Fig. 4). At the side of some long ranging species occurring over several zones (7, 8, 9, 11) each zone is characterized by a certain number of species which are restricted to it. One of them is chosen as index species and gives its name to the zone.

But a chronological distribution of species of this kind is highly improbable. Even if we assume that successive species (e.g. 1 and 2) are linked phylogenetically, it is statistically impossible that a change on species level should occur at exactly the same time in independent lineages. A range chart of this type can only be obtained if there are gaps between successive zones.

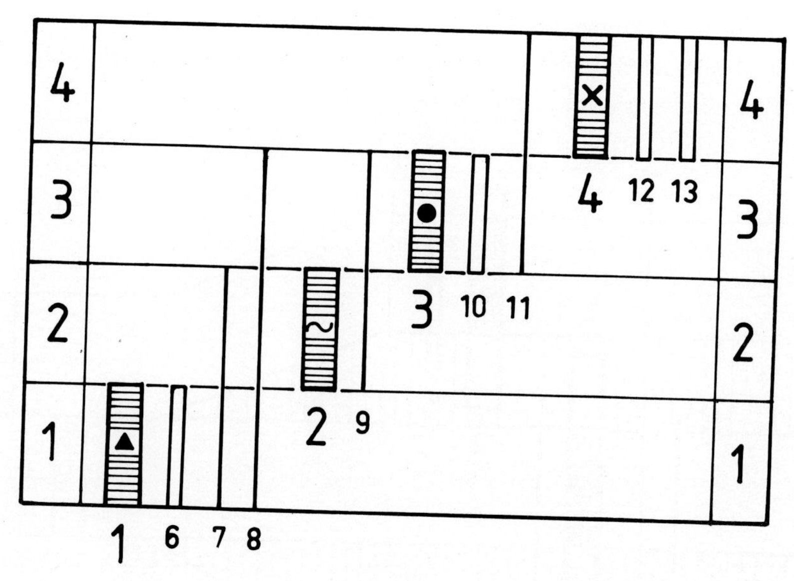

Fig. 4

Scheme showing what the chronological distribution of species would be, if all biochronologic zones of a zonation would be total range zones, each corresponding to the life time of its index species (further explanation in the text).

A realistic range chart based on a complete documentation would look like this (Fig. 5). None of the phyletic events — first appearances and extinctions of species — would be exactly isochronous with another one. But if we introduce a certain number of gaps, we obtain successive zones each corresponding to the life time of an index species.

(2) In the case of Oppel's zones we probably deal mostly with barren intervals instead of stratigraphic gaps. In any case, Oppel-zones are assemblage zones defined by their contents but without clear cut boundaries and with indeterminate intervals between them. Discrete zones of this type provide nevertheless a powerful tool for correlation if all species of the assemblage are short-lived. This is indeed the case with ammonites and still today ammonite stratigraphers work preferably with this type of zone: A zone is characterized by a bundle of species having a similar range and being excluded from the next assemblage of this kind. A given fauna can then be placed in a zone even if only a part of its characteristic assemblage has been found. The advantage of this method is obvious, but there are also serious drawbacks: many associations will be difficult to interpret. Interpretations will vary from one author to another and, in the absence of clearly defined boundaries, zonal attributions remain to some degree arbitrary. At least they are difficult to control in critical cases.

(3) Therefore the Oppelian approach should always be coupled with some method of quantitative stratigraphy — unitary associations in the sense of Guex[5] are the most appropriate.

It should also be remembered that zones without strictly defined boundaries are incompatible with the GSSP concept, i.e. with boundary stratotypes. As we have seen, the definition of a chronostratigraphic boundary by means of a GSSP has to be preceded by careful choice of a marker level, which will mostly be a biostratigraphic level supposed to be close to its biochronologic equivalent.

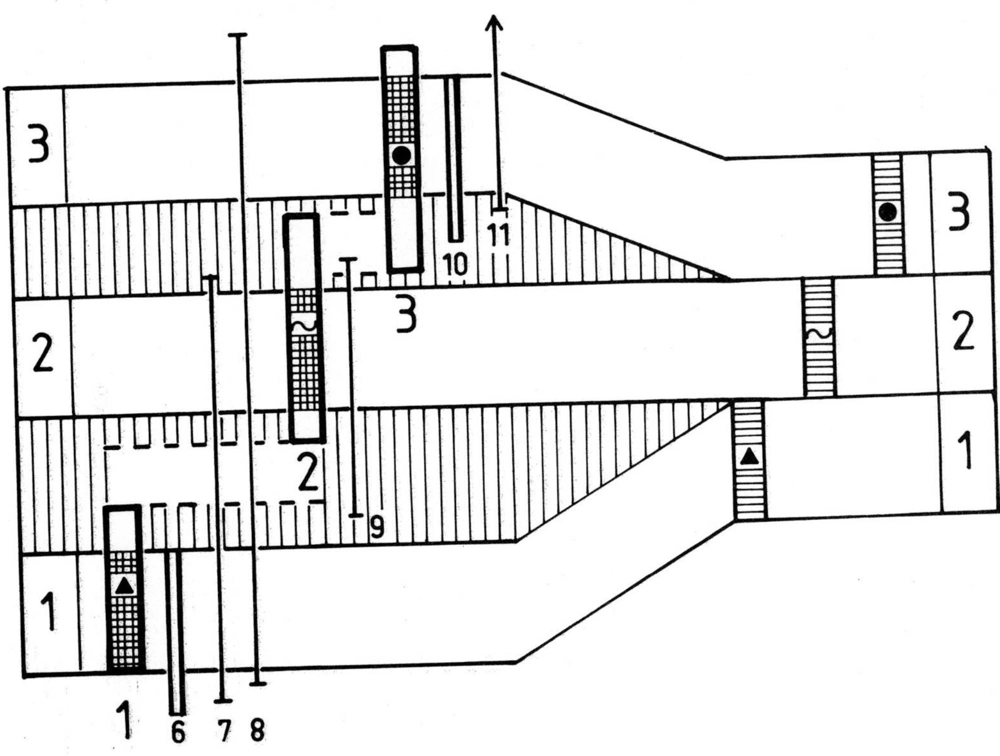

Fig. 5

The same zonation as in Fig. 4, but this time the range chart is based on the realistic assumption that two events (first and last appearances of species) will never occur at exactly the same time. The zonation of Fig. 4 (at the right) can, however, easily be obtained, if we assume that there are no biostratigraphic data from intervals between the zones (either barren intervals or stratigraphic gaps).

So our next subject will be what kind of biochronologic zone concept best fits the GSSP concept.

2. Biochronologic zones with strictly defined boundaries

1) Illustration of the problem

Hedberg (1976, Figs. 4-11)[2] gives an overview of different types of biostratigraphic zones which applies as well to biochronologic zones. It appears that the different types of assemblage zones have no unambiguous boundaries: the boundaries of the multitaxon concurrent range zone (Hedberg, 1976, Fig. 6)[2] are overdefined — as stated above, it is practically impossible that several species appear exactly at the same time. In the assemblage zone and the Oppel-zone (Figs. 4, 8, both are in reality identical) there are no clear cut boundaries but transitional intervals. Consequently there are no definable units — unless we choose one event for each boundary — but then we are in the group of concurrent range zones, etc. to be discussed below. Acme zones are not clearly definable either.

So we come automatically to the following conclusion: a zonal boundary will only be clear cut and unambiguous if it is defined by one single evolutionary event. "Evolutionary event" means then phyletic first appearance or extinction of a species.

2) Different types of biochronologic zones

(1) If we follow the rule "one boundary one event", only five types of zones are possible, corresponding to the five possible combinations of first appearance and extinction (Fig. 6).

(a) We may have a taxon-range zone = total range zone, reaching from the first appearance to the extinction of one and the same species.

The 4 other types of zones use the combination of 2 species. With respect to them three of them are partial range zones.

(b)+(c) Consecutive range zones, are based on consecutive first appearances or extinctions. With regard to their contents they are characterized by the presence of one index species and the simultaneous absence of the other one.

(d) Concurrent range zone: this is the time interval of co-occurrence of two index species.

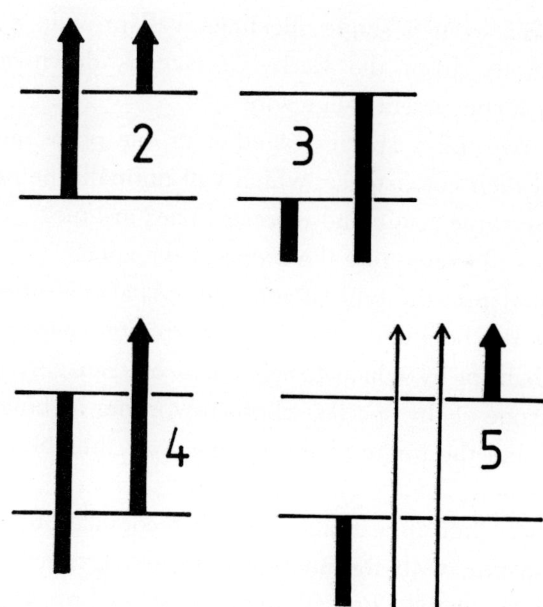

Fig. 6

If we follow the rule: "one boundary-one event", only 5 types of biochronologic zones are possible: 1. Total range zone (not figured); 2, 3. Consecutive range zones; 4. Concurrent range zone; 5. Interval zone (for further explanations see text).

(e) Interval zone corresponding to the interval between the extinction of the first index and the appearance of the second one — thus characterized by the absence of both index species.

(2) As to the practical value of these five zone types, the interval zone is without any doubt the least recommendable one, because it is only negatively defined with regard to its contents.

In this respect, the total range zone is superior to all other zone types, since it combines a clear definition by its contents with an unmistakable boundary definition. Unfortunately — as shown above — it is highly improbable that a succession of several total range zones will occur in nature. Hence the majority of the zones forming a biochronological scale will be concurrent or consecutive range zones, i.e. teilzones which do not cover the whole life span of their index.

As to the recognition of zones, the primary elements are of course given by the definition, i.e. the type of zone. But other species may play an important role in the recognition of the zone. As a matter of fact, whatever the definition of a zone may be in its strict sense, it can be delimited in a way that it matches closely with a characteristic assemblage. This is very important because the obvious advantage of assemblage zones may thus be combined with the postulate of a strict boundary definition.

In a general way, we may say that the recognition of zones is a matter of approximation. We know that biostratigraphic field data, i.e. the local range of a given species will often not represent its whole life span. But if we deal with short lived species of a life time of 1-2Ma, even a single specimen will provide a more precise datation than radiometric methods, from the Early Cretaceous downwards. The same is true (in another way) for a concurrent range zone.

Finding the two index species together in one place means that we are somewhere in the interval of their coexistence, which will normally be of very short duration, too. In this respect consecutive range and interval zones are more delicate. Here, imperfect biostratigraphic data will exaggerate the scope of the zone.

Independently from the type of zone there is also a difference in reliability according to the nature of the boundary. Except bioturbation, which is always of limited extent, there is no mechanism by which fossils could be transported into strata which are older than the chronozone of the species. Slight reworking is, however, often difficult to recognize in the field and the range of the species may thus be extended beyond the limits of its existence.

In some cases, first appearances can be recognized by the existence of intermediate forms, forming a link with the ancestral species. Extinction, on the other hand, is not morphologically distinctive. For all these reasons, first appearances provide in general more reliable zonal boundaries than extinctions.

3. From biostratigraphy to biochronology

1) Nature of the problem

Up to now the relations between biostratigraphy and biochronology were treated as if a definite zonation was already available. The only problem was then, to which degree of precision these ideal zones could be determined using incomplete biostratigraphic field data. The aim was to show, how and how far this determination depends on the nature of stratigraphic boundaries (first appearances or extinctions) and the underlying zone type. The problem we have to study now is by which methods we can establish a reliable zonation, i.e. a zonal scheme which corresponds to the real succession of evolutionary events.

In some cases, especially if we deal with microplankton in pelagic facies, there is always a certain number of series providing a complete and unbiased record of the chronologic succession of species. In this case the only problem is one of classification : how to define a workable boundary, how to create discrete units in a continuum.

But very often—and always in the case of macrofossils such as cephalopods—we have to reconstitute the succession of evolutionary events from more or less scanty data. The resulting zonation is no longer the direct expression of observed fact (as in the case of microplankton). A large part of interpretation or extrapolation—or if you prefer—abstraction intervenes. There is a great danger of premature generalization and circular reasoning. That is why so many zonations did not stand the test of practice and ran into contradictions.

2) Biochronologic interpretation of biostratigraphic field data

To introduce the problem I come back to Fig. 1. Here the problem is simplified, it is not the succession of phyletic events but of whole species which is considered. But the example illustrates very well how different sections may complete each other mutually, so that it is possible to reconstitute the entire succession of species. This succession corresponds to a synthetic view, it is an abstraction because it is more than what can be observed in any single profile. But nevertheless it is the expression of historic reality. On the other hand, as illustrated by Figs. 4 and 5, a mere succession of species is not a sufficient base to establish a suite of contiguous zones.

The next step is therefore to reconstitute the correct succession of phyletic events, especially when the ranges of species overlap. The problem is to find out whether the observed biostratigraphic succession reflects historical reality, whether it is biogeographical or whether it is biased by biogeographical or one of the other factors mentioned above. This problem becomes the more acute, the closer the events under question are.

It is obvious that two events which are very far apart, let's say 10Ma or more, will always appear in the same biostratigraphic order, so that there can be no doubt about their true chronological succession. But if we want to arrive at a finer subdivision, we run soon into difficulties. In comparing different sections we obtain contradictory biostratigraphic data. Unless we are able to distinguish between primitive and advanced variants of a species, there is no possibility to decide which succession gives the true version.

Under these circumstances, two approaches are possible, a qualitative one and a quantitative one. Both are approximations in the sense that the proposed zonation is not absolutely certain but reflects the most probable succession of events.

The qualitative approach corresponds to the classical alternative, it is selective. Those species are chosen as guide fossils, which show the least contradictory biostratigraphic distribution. This will normally be species which are short lived, not very sensitive to facies changes and have a wide geographic range. For a long time these properties have been postulated as prerequisites for good guide fossils. But in any case it is very important to obtain statistically significant results, even though the method is purely qualitative.

Once the chronological succession of selected phyletic events can be supposed to be correctly determined, we can use these events to define zonal boundaries. Theoretically, all of them may be made zonal boundaries, but one may also prefer to have some of them as subzonal boundaries only. This decision is a matter of practical consideration.

There are, of course, limits to the precision we can attain. It seems that the maximum resolution which can be obtained on a regional scale is in the order of some 100 000 a.

3) Quantitative stratigraphy

(1) There are different methods of quantitative stratigraphy. They have in common that ideally all observed species are used for chronocorrelation. So these methods are not selective, at least as long as the number of species is not beyond the capacity of the available computer. In any case the number of species taken into consideration is considerably higher than in the classical selective approach described above. Three of these methods will be discussed very briefly, they were initiated by Shaw,[6] by Hay[7] — this one further developed by Gradstein and Agterberg[8]— and by Guex.[5,9] For a critical somparison see Baumgartner.[10]

(2) The principal aim of Shaw[6] and of Hay/Gradstein and Agterberg is not to establish zonations in the classical meaning of the term. Both are in the first place methods of chronocorrelation based on a statistical analysis of biostratigraphic data. The underlying philosophy is the same in both approaches : biostratigraphic field data are inevitably contradictory as to the succession of phyletic events. A statistical evaluation of data should allow finding out which is the most probable succession of events. The analysis is based on the assumption that observed first and last occurrences will show a random distribution centered around the real phyletic event. This premise is wrong. An observed first occurrence cannot be prior to the phyletic appearance of the species. In the same way, its last occurrence cannot be after to its extinction, unless we deal with reworked material. In other words, probabilistic methods (as they were called) reduce the ranges of species artificially and marginal overlaps will not be recognized. I believe nevertheless that reasonably accurate correlation can be obtained, as the same systematic error is introduced everywhere.

(3) The approach proposed by Guex[5,9] is deterministic (for an explanation of the method in English, see Baumgartner[10]). The difference with probabilistic methods appears, when we consider the following case : in 99 sections two species succeed each other and in the 100th profile they occur together. In probabilistic stratigraphy this case would not be taken into account, being statistically insignificant. One would continue to consider the two species as successive and mutually exclusive, which is obviously wrong. In the deterministic method all co-occurrences are retained and plotted in a matrix. Taking also into account local superpositions, so called virtual co-occurrences can be established: if species A appears below B in one section and above B in another, it is obvious

that A and B must be at least partially coeval. It would lead too far to go into the mathematical details of this method, like application of Graph Theory, etc. The final results are "Unitary Associations" (UA) which correspond to successive sets of the maximum number of co-occurring or potentially co-occurring species.[10] UAs are a kind of biochronological zone: like Oppel zones, they are discrete zones. They provide, however, a more detailed subdivision than the classical, intuitive approach because the number of species on which they are based is considerably higher. The mathematical approach has the additional advantage that the procedure can be better controlled than in the case of an intuitive zonation.

On the other hand, a scheme of UAs may be modified by new data, like all other kinds of biochronological zonations derived from fragmentary biostratigraphic data. In the same way they correspond to synthetic schemes which may very well not be present in any one single section (cf. Fig. 1). They provide, however, a standard of reference to express the relative age of a fauna in an accurate and flexible manner.

CONCLUSIONS

1. Introduction

To conclude with I shall recall here the most important points from the methodological point of view, starting from the construction of a biochronologic scale on the base of imperfect biostratigraphic raw data.

The important point is that in any case we have to find out the correct or at least the most probable order of phyletic (or evolutionary) events, i.e. first appearances and extinctions. In other words we have to establish a synthetic biochronologic range chart. This is completely independent of the question if we want to define zones or not.

2. Definition of biochronologic zones

1) General premises

As we have seen, we have to follow the rule "one boundary- one event", if we want to arrive at zonal definitions which are unambiguous. Biochronologic zones are defined by evolutionary events which often will not be directly observable. In this case they correspond to a theoretical concept even though they reflect historical reality.

2) Technical aspects

I need not come back to the five possible types of zones described above. Very often they are of a more theoretical importance. When the zone is made to match closely with a characteristic assemblage, the distinction between a concurrent range zone and a consecutive range zone will be of lesser importance. Their only function is to provide clear cut boundaries.

It has also to be stressed that biochronologic zones should be defined by their lower boundary only, the upper boundary being given automatically by the lower boundary of the succeeding zone.

3. Recognition of biochronologic zones

Normally a zonation will be established in successions which are particularly rich in fossils. They will often have to be applied, i.e. recognized under less favourable conditions. Biostratigraphic data may only allow more or less rough approximations to the ideal scheme.

But here I may recall the example given above, showing how we can combine biostratigraphic and lithostratigraphic correlation techniques in order to determine the stratigraphic position of biochronologic boundaries.

REFERENCES

1. Cowie JW, Ziegler W, Boucot AJ, Bassett MG and Remane J, Cour Forsch -Inst Senckenberg, 83 (1986), 1.
2. Hedberg HD (ed), International Stratigraphic Guide, J Wiley and Sons (1976), New York.
3. Oppel A, Die Juraformation Englands, Frankreich und des sudwestlichen Deutschlands, Ebner Seubert (1856-1858), Stuttgart.
4. Harland WB, Cox AV, Llewellyn PG, Pickton CAG, Smith AG and Walters R, A Geologic Time Scale, Cambridge Univ Press (1982), Cambridge.
5. Guex J, Correlatios biochronologiques et Asslciations unitaires, Presses polytechn. romandes, Lausanne (1987).
6. Shaw AB, Time in Stratigraphy, Mc Grawhill, (1964), New York.
7. Hay WW, Eclogae geol Helv, 65 (1972), 255.
8. Gradstein FM, Agterberg FP, Brower JC and Schwarzacher WS, Quanrtitative Stratigraphy, D Reidel Publ Comp/ Unesco (1985), Paris.
9. Guex J, Bull Soc vand Sci nat, 73 (1977), 309.
10. Baumgartner P, Computer and Geosci, 10 (1984), 167.

XIe Congrès International de Stratigraphie et de Géologie du Carbonifère Beijing 1987,
Compte Rendu 1 (1991): 201-204

ON MOSCOVIAN STAGE STRATIGRAPHY: PROGRESS IN STUDIES

M. N. Solovieva, V. S. Gubareva, N. V. Goreva, O. A. Betekhtina,
O. P. Fisunenko, V. K. Teteryuk, A. V. Popov and I. I. Dalmatskaya
(The U.S.S.R. National Committee of Geologists, 109017 Moscow, Pyzhevsky per,. 7)

Considerable progress in the study of Moscovian Stage is attributed to the results obtained from repeated studies of interstage divisions in stratotype territory of Moscovian Stage, which makes it possible to have a better understanding of the volumes of suites, horizons and stage and offers a new zonal scheme: (1) Vereisky horizon, corresponding to the zones of *Profusulinella cavis, Aljutovella aljutovica, Al. artificialis;* (2) Tsninsky horizon, corresponding to the zones of *Aljutovella priscoidea, Al. znensis, Hemifusulina volgnesis;* (3) Kashirsky horizon, corresponding to the zones of *Hemifusulina kashirica, H. moelleri, Beedeina pseudoelegans; Moellerites lopasniensis, Beedeina ozawai, Fusulinella subpulchra;* (4) Podolsky horizon, corresponding to the zones of *Beedeina elegans, Fusulinella colaniae; Fusulinella vozhgalensis, Fusulina ulitinensis; Beedeina kamensis, Putrella brazhnikovae;* (5) Myachkovsky horizon, corresponding to the zones of *Fusulinella bocki, F. rara, Beedeina samarica, Fusulinella podolskensis, Fusulina cylindrica domodedovi.*

Making the conodont scheme more precise resulted in introducing into the Moscovian Stage the zonal scheme double ("parallel") zones in accordance with the succession of the species of *Streptognathodus* applicable to more subsided areas of the basin and the subdivisions according to the succession of the species of *Neognathodus* (for shallow water deposits). Vereisky horizon corresponds to the Zones *Streptognathodus transitivus* and *Idiognathodus fossatus.* The chacateristics of Tsninsky horigon need to be more specified. Kashired horizon connects with the *Streptognathodus dissectus* Zone and the *Neognathodus bothrops, Neognathodus medadultimus* Zones respectively. Podolsky horizon is correlated with the *Streptognathodus concinnus-Idiognathodus podolskensis* Zone that meets with the zone of *Neognathodus medexultimus.* Myachkovsky horizon corresponds to *Streptognathodus cancellosus* Zone, which meets with the *Neognathodus roundyi* Zone of the parallel scheme. Based on the standard sequence of Moscovian Stage, changes were introduced practically into all regional schemes, forming a part of tropical paleoclimatic band (belt) and a new correlation model. Sequences of central and eastern regions of the East European Platform were divided and correlated in accordance with the stratigraphic units (divisions).

Thus, analogues of Tsninsky horizon are widespread to the east from the stratotype region. A patch of dolomites with clay and limestone partings can be traced, starting from the Tokmovsk elevation and its spurs, along the key sections of stratigraphic wells such as Tokmovo, Alatyr, Sundyr, Zubova Polyana, Prudy, Baranovka and Kikino upon Vereisky horizon. In its basement one can observe pebbles of Vereisky horizon rocks, and in the top partings of conglomerate, the development of which is the evidence, if not of continental break, then, at least of a considerable shallowing. This patch includes aljutovellic-*Profusulinella* foraminifer complex (the first patch—by I.I. Dalmatxkaya, 1954) and brachiopods of *Choristites priscus* group. Its thickness is 15-20m. Along the outlying districts of Ural-Povolgye (Oparino, Sovetsk, Shikhovo-Chepetzk), the horizon is of a binary building: 7-13-18m of terrigene motley rocks in the lower part, and a part of carbonates on top, which are characterized by zninkashira foraminifer complex. Along the outlying districts of Ural-Povolgye (Oparino, Sovetsk, Shikhovo-Chepetzk), the horizon is of a binary building: 7-13-18m of terrigene motley rocks in the lower part, and a part of carbonates on top, which are characterized by zninkashira foraminifer complex.

Along the western outlying districts of Ural-Povolgye (wells: Syzran, Ulyanovsk, Melekess, Orekhovka), Tsninsky horizon is represented by a series of organic-detritus and organic-detrital limestones with dolomite and clay partings; with a thickness of 25-30m. As a rule, at the lowest 18-25m is developed an interbed of calcareous conglomerate, above which no terrigene material admixture can be observed. The complex is characterized by foraminifers: *Ozawainella digfitalis, Schubertella gracilis, Aljutovella priscoidea, Al. postaljutovica, Al. znensis, Profusulinella librovitchi, Pr. prisca, Pr. timanica, Pr. sphaeroidea, Pr. ovata, Pr. constans, Eofusulina triangula* and choristites, belonging to the *Choristitus priscus* group. Faunistic characteristics of the series corresponds to the same one of Tsninsky horizon.

In the central part of Ural-Povolgye, beginning from Buzuluk in the west and up to Orenburg in the east, Tsninsky horizon is also observed. In the first case its deposits consist of sufficiently clayey limestones and dolomites, and organic-detritus detrital limestones in the second, which was distinguished by I.I. Dalmatseaya as a local subzone *Aljutovella priscoidea.*

To sum up, deposits of this subzone include typical complexes of Tsninsky horizon fauna. In southern Orenburgye the whole composition of Moscovian scale is represented by carbon-bearing rocks, but in a number of sequences Tsninsky horizon includes carbon-bearing rock debris. In such sequences its thickness is less in size, and higher up along the series an interval is fixed, in the absence partly or completely of Kashirsky and sometimes also partly of Podolsky horizon. Deposits of Tsninsky and Kashirsky horizons in this region vary from 70-90m to 20-60m in thickness.

Thus, Tsninsky horizon, distinguished according to its stratotype in an exposure in Oksk-Znin rampart, can be observed practically everywhere in the central and eastern regions of the Russian Plate.

In sequences along the southern wall of the Moscow syneclise and at Tokmov vault, lithological composition and thickness (12-18m) of Tsninsky horizon approach to the values given above for the stratotype (compared with Well 1 Mosolovo, Well 1 Zubova Polyana, Well 1 Tokmovo, Well 1 Prudy, Well 1 Poretzkoye, Well 1 Alatyr).

In a transitional zone from the central to the eastern regions, traces of shallowing or development of sub-continental conditions are reflected in a form of conglomerate or palygorskite clay partings. In the western regions of the south wall of the Moscow syneclise it resulted in the formation of Polustovogorsk Series, which is a continental analogue of Tsninsky horizon. Simultaneously, it can be observed that the thickness of Tsninsky horizon is up to 20-25m (Kikino, Syzran).

Farther to the east: on the left bank of the Volga River and in Zavolgye, carbon-bearing formations become predominant in Tsninsky horizon. Unstable conditions of sedimentation are reflected in the development of series of carbonaceous breccia in the foundation and considerable supply of quartz-feldspar material. The thickness of deposits reach up to 30-45m.

Having considerably incomplete faunistic characteristics of deposits in the sequences of wells, in the patches of rocks observed, which were distinguished into Tsninsky horizon, it, to some extent, meets the stratotype complex. It was stated then, that foraminifer complexes reflect the same succession change of the associations, which was stated in a standard sequence of Moscovian Stage (Solovieva, 1984, 1987).

Brachiopods of Tsninsky horizon from sequences described above include a small group of forms, the most typical of which is *Choristites priscus*.

On the eastern part of Pricaspian depression, in Aktyubinsk Priuralye, analogues of Tsninsky horizon can be observed in a number of sequence. Thus, according to I.I. Dalmatskaya definitions of foraminifer complexes of Vereisky, Tsninsky, Kashirsky, Podolsky and Myachkovsky horizons are distinguished in the sequences of wells Alibekmola and Zhagabulak.

M. N. Solovieva (1987) introduces for Middle Asia two new regional horizons, Nurataussky, correlated with Tsninsky horizon, and Yettysaisky, with Kashirsky horizon of the Russian Plate.

Besides, new data are obtained, concerning the character of comparison of Moscow deposits with the zonation scale of Western Europe, and relations to schemes of various biogeographical areas.

These data serve as a confirmation of conclusions on cephalopods for the first time drawn by A.V. Popov, which concern the conformity of the lower boundary of Moscovian Stage up to WA-WB, i.e. O.A. Betekhtina, while studying non-marine myrians, established approximate conformity of the lower boundary of the stage with the level of the boundary WA-WB, and the upper one, with the foundation SA_2. Thus, the lower boundary is traced, according to Donbass scheme, on limestone level i_3^1, while the upper one on N_1, and for Angarida, the lower boundary meets the foot of Mazur suite, while the upper one the top of Alakayev suite.

According to V.K. Teteryuk, there is a rapid change of microspore complexes, which are typical for Moscovian fossil succession of Bashkirian ones, in Donbass sequences they are established near the level of limestone K_1, which is rather close (level K_1)

to the data on cephalopods, non-marine myrians, conodonts and foraminifers.

Based on foliated flora, according to O.P. Fisunenko, the top of suite C_3^2 and the base of suits C_2^4 of Donbass, meet the upper part WA of Western Europe. So, one can possibly suggest that the boundary of Bashkirian and Moscovian stages might be established near the foot of suite C_2^5, and if correlated with the foundation of the upper patch of upper Asovian Series of Podmoskovye, also meets very closely limit WA-WB.

XIe Congrès International de Stratigraphie et de Géologie du Carbonifère Beijing 1987,
Compte Rendu 1 (1991): 205-212

APPLICATIONS OF PALAEOMAGNETISM IN THE CARBONIFEROUS

D. H. Tarling
(Plymouth Polytechnic, Plymouth PL4 8AA, England)

ABSTRACT

Palaeomagnetism can be used for dating on three different time scales, although only the geomagnetic reversal and polar wander scales are directly applicable to Carboniferous times. The current knowledge of Carboniferous magnetostratigraphy is inadequate, with evidence of remagnetization and regional heating affecting the most recent studies. Polar wander path dating is becoming better established for the major blocks and offers high precision dating for independently moving blocks. The structural applications of palaeomagnetism have still been largely concerned with establishing the palaeogeography, particularly the palaeolatitude, of the main tectonic units. Since the last IGCC, the reconstruction for Gondwanaland has been improved, but the main development has been better palaeomagnetic control on tectonic units within eastern Asia. Other palaeomagnetic applications, such as determination of the depth of burial and time of uplift, are possible, although such applications are not specific to the Carboniferous.

INTRODUCTION

The magnetic properties of rocks are readily determined in most instances and have a wide range of possible uses within geology. The two main applications are the use of magnetic remanence for dating and structural analyses. In addition the magnetic mineralogy is a good indicator of the redox conditions to which rocks have been subjected during their history. The remanent magnetic properties can also be used to determine the past chemical changes induced by temperature, as well as using the remanence to estimate thermally induced remanence such as that associated with burial and subsequent uplift — the date of which can also be defined by palaeomagnetic dating.

The basic principles of palaeomagnetism are now well established,[1,2] although there are still many fundamental problems to be resolved. Similarly the instrumentation required for adequate palaeomagnetic analyses have been described elsewhere.[3] Neither of these aspects will be considered in this article which only considers the applications of palaeomagnetism with particular reference to the Upper Palaeozoic.

PALAEOMAGNETIC DATING

Palaeomagnetic dating can be applied on three main time-scales.

1. Secular variation of the geomagnetic field, that has periodicities of the order of 10^2 -10^4 years during Recent geological time, but the behaviour of the geomagnetic field, both now and in the past, is not predicatable and absolute dating by this method relies on establishing a dated record of the secular variations as preserved in fired or sedimented materials. As there are no scientific dating methods of adequate precision for establishing such scales during the Upper Palaeozoic, such absolute dating is impractical for such times. However, it may be possible to either (a) assume secular variations of similar periodicities as at present, or (b) dating can be acheived by assuming specific rates of deposition of eruption. Relative dating is, of course, possible using such techniques, and it is possible that secular variation patterns, preserved in either igneous or sedimentary sequences, could be used to match against those found in other sequences elsewhere. However, experience suggests that all such techniques will be invalidated because of the prevalence of subsequent chemical or thermal events affecting such rocks. The only areas where such small time-scale techniques can be applied is where palaeothermal indicators, such as conodont colouration, vitrinite reflectance, hydrocarbon maturity, etc., indicate that there has been little or no regional heating or chemical activity. The Chinese Tarim block, for example, may afford a region where original Carboniferous secular variation patterns may be preserved.

2. Geomagnetic reversals of polarity during the Tertiary take some 3 000 -10 000 years to complete and there are no reasons for considering that significantly different time-scales are involved during the Upper Palaeozoic. Such polarity changes thus form global, geologically instantaneous time markers that are ideal for correlating biostratigraphical and radiometric time scales throughout the world. However, while the Mesozoic-Cenozoic sequence of polarity changes is preserved in the ocean-floor anomaly patterns, the situation for earlier times is much less clear as such polarity change sequences can only be established in sedimentary and igneous seqnences, the age of which must be obtained biostratigraphically or radiometrically. Unfortunately, most sedimentary and volcanic sequences are discontinuous in space and time and there are fundamental problems in distinguishing between the magnetic age of the polarity changes as these do not necessarily correspond to the age of the rocks themselves. Igneous rocks are most likely to have a magnetization associated with their original cooling, but consolidated sediments will normally have acquired their stable remanence during diagenetic processes which normally occur a few 10^2 to 10^4 years after deposition. However, such changes may be delayed for several 10^6 years. However, even the stable remanence may well be associated with even later thermal or chemical changes, such as those associated with burial and later unroofing.

Few attempts have been made to establish the magnetostratigraphic column for the entire Carboniferous, although some of the earlier studies were particularly concerned with establishing the time of initiation of the Kiaman reversed polarity interval that persisted throughout the Permian and possibly into the Upper Carboniferous.[4] The latest review of the Carboniferous polarity time-scale[5] attempted a compilation of previously published observation on the polarity scale for the various continents, together with the results of two specific studies of type sections in the British Isles comprising the Upper Carboniferous[6] and Lower Carboniferous.[7] These records have been incorporated in Figure 1, with a modification to include the latest reports of the polarity sequences determined in the U.S.S.R.[8] However, while the British data have been retained in this compilation, this has been done to demonstrate some of the difficulties in establishing such a zonation for the Carboniferous, or for any pre-Jurassic Period.

The British Upper Carboniferous was established using type sections within England and Wales.[6] At that time, several of the fold tests were shown to indicate that the remanence was commonly associated with the folding process, i.e. the closest grouping of vectors occurred when the folding was partially unfolded. However, the age of the folding was not clear. Some folding was probably penecontemporaneous with deposition as the sediments on the carbonate platforms were draped over faults that appear to have acted as distinct facies barriers at this time. However, many of these fold structures were either enhanced or originated in association with the Variscan orogeny and may not have formed until Early Permian times. The age of the magnetization is therefore uncertain for this period. In the Lower Carboniferous, type sections were examined in southern Ireland[7] but as these are mostly defined near to the Variscan front, they showed clear signs of metamorphism, so equivalent sections further north were sampled in areas north of the Variscan front where it was considered that the largely flat-lying rocks were unaffected by Variscan events. The presence of consistent polarity horizons suggested that the magnetization was primary, but no standard fold-tests could be undertaken because of the shallow tilt of the beds. However, the mean inclination for these sediments suggested a palaeolatitude of some 11^0 N—which is that expected for Early Permian times, and inconsistent with the low, southerly palaeolatitude predicted by interpolation between Upper Carboniferous and Devonian European observations.[1,9] It has subsequently been shown that the sampled localities have been subjected to unexpectedly high temperatures, probably $\geqslant$200-250^{0}C, on the evidence of conodont colouration (Sevastopoolos, G. pers. comm.). As this temperature, if held for some 10 million years, is equivalent to laboratory heating to$>$450^{0}C. So that most, possibly all of the original remanence has been replaced by the magnetization acquired on cooling from such burial temperatures, probably in the Lower Permian. It is therefore not clear how much reliance can be placed on the British data. However, it must be emphasized that the British data were specifically planned to resolve the magnetostratigrphy and specific tests were undertaken to establish the age of remanence. While the data from the U.S.S.R. were also specifically examined for magnetostratigraphical properties, no testable information has been provided on the age reliability of such observation. For other parts of the world, there are generally no fundamental controls on the age of remanence.

Further investigations are being planned into this problem in both Ireland and Britain, but it is clear that such magnetostratigraphic studies must be conducted in areas where there are no indications of burial, tectonic or thermal activity. When considering the optimum location for establishing a magnetostratigraphic sequence, as being considered for China, it is clearly important to select areas where that have not been subjected to later heating, i.e. the conodonts, spores, etc., should not be discoloured, any coals in the sequence should be of low rank, etc. It is also necessary that the age of the remanence can be established, such as by fold tests, although this requirement unfortunately also means that a degree of flexure is required with the inherent possibility of remagnetization being associated with the folding process itself.

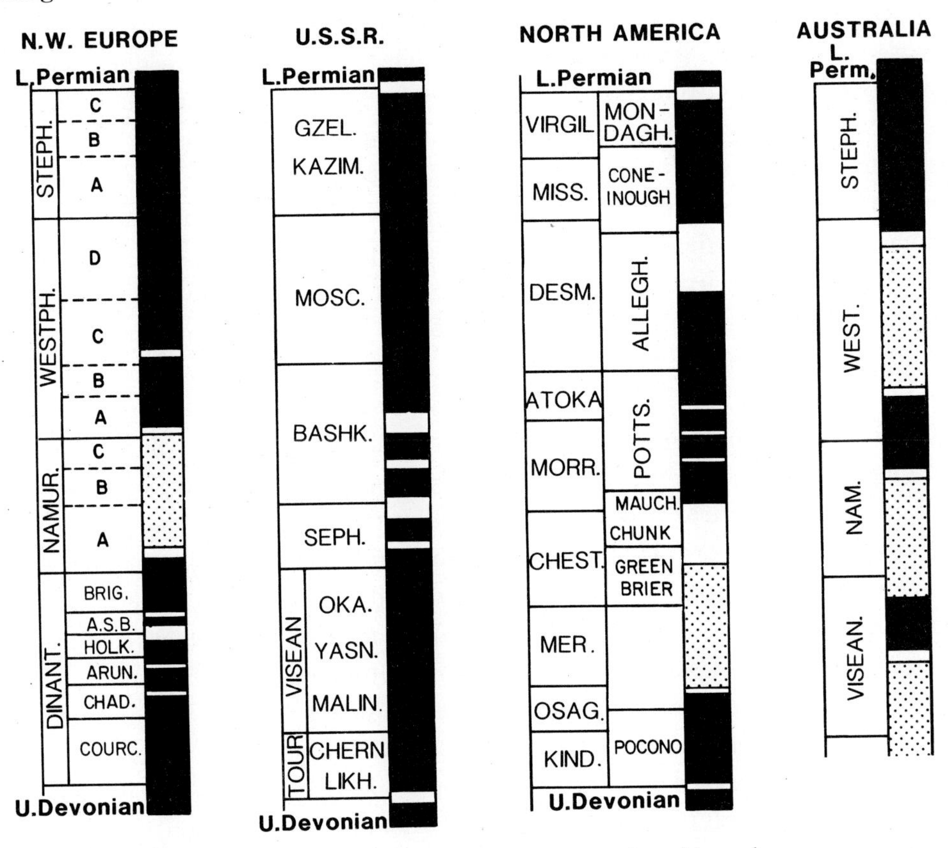

Fig. 1 Some Carboniferous magnetostratigraphic scales

(modified after reference 5, with particular changes to the USSR column[8])

Nonetheless, while considerable difficulties exist in establishing such magnetostratigraphic scales, the significance of such precise time horizons on a global scale is of fundamental importance in establishing the correlations between different biostratigraphic techniques (particularly between the Laurentian and Gondwanan zonations), the verification of eustatic models for sea-level changes during the Carboniferous (with associated palaeoclimatic implications), and so on. At the moment, the brevity and global character of polarity changes of the geomagnetic field makes magnetostratigraphy a uniquely applicable tool for testing such fundamentally important models. The fact that there are difficulties in establishing the scale, as indicated above, should not detract from the essential need to establish such a zonation in each of the supercontinents. It should also be commented that the local failure to establish such a scale can also be of importance in evaluating the burial or thermal history of specific areas—such as now to be investigated for the lead-zinc deposits in southern Ireland.

3. An apparent polar wander path is created by the movement of a tectonic unit, relative to the average geomagnetic pole. Such paths can be used for dating the remanence of other strata from the same tectonic unit. In this technique, the direction of remanence of unknown age is used to calculate the corresponding pole position, using standard formulae, and this pole position can then be dated from its location on the dated apparent polar wander path. For the Laurentian continents, the average rate of motion during the Mesozoic-Cenozoic is some 0.3^0 per million years. If a pole position is calculated to some $+3^0$, then the average dating accuracy corresponds to $\pm$ 10 million years. In general, small tectonic units, when not linked to a supercontinent, have more rapid motion relative to the pole. The current research to establish the polar wander paths for each of the Chinese tectonic units [10-14] should therefore provide a more precise dating of the remanence than is possible for the larger, more slowly moving continents. It must be emphasized, however, that this method dates the age of the magnetization, which may be the same or younger than the geological age of the rocks themselves.

This magnetic dating method can be used for relative dating if the polar wander curve has not yet been established for the time or area of interest. In such a context, the syngenetic deposits on the same tectonic plate will have similar pole positions. This allows syngenetic ore bodies to be distinguished from epigenetic deposits for which the ages of magnetization will differ and hence their corresponding pole position will differ. Such conclusions can be draw even if the actual ages are individually indeterminate until the polar wander curve has been established.

STRUCTURAL

The use of palaeomagnetic methods for determining the past configuration of continental blocks is well known, but it has not be so commonly realized that identical arguments can be used to determine the past relationships between small tectonic units. The palaeomagnetic study of "exotic" terrains in the Western Cordillera provides one example on a scale of several thousand square kilometers,[1] but palaeomagnetic directions can be rotated on all scales, down to that of individual fault blocks of some 200×50kms, or even scree slopes and soil creep on the scale of a few square meters.[1]

Some of the problems of forming palaeomagnetic reconstructions for the Carboniferous Period were outlined previously[9,15] in which it was shown that major features, such as the existence, or not, of a Mid-European Ocean was still not determinable from the extant palaeomagnetic data. There has been little fundamental change in the palaeomagnetic data base since that time, with two particular exceptions:

1. While the "final" reconstruction of Gondwanaland has probably not yet been

acheived, it seems unlikely that there will be further major changes in the reconstruction for the major continental units of this supercontinent based on both ocean-floor magnetic anomaly pattern matching and continental palaeomagnetic data.[16] As the main continental components of this supercontinent appear to have existed through most of the Palaeozoic, such a reconstruction can be used to extrapolate palaeomagnetic observations between these major units (Fig. 2). There are, however, still uncertainties about the changing relationships of these major units with their marginal areas.

2. The high quality of recent palaeomagnetic work in the Far East[10-14] now allows a better evaluation of the palaeogeographic relationships within the eastern Tethyan region (Figs. 2,3), including the Carboniferous Period.

It must be emphasized that these reconstructions are still uncertain, within at least $\pm 5\text{-}10^0$ of latitude for even the major continental blocks, and there is little or no data yet available for the precise location of present-day "exotic" terrains, such as those now forming the Pacific margin of the America. This means that the indication of a Pacific resource for such areas is speculative and there are not yet any palaeomagnetic data relevant to this problem. Nonetheless, it is thought that the Carboniferous reconstructions are now becoming sufficiently reliable to attempt reconstructions of possible oceanic circulation patterns, with the intention of comparing the associated climatic patterns in terms of extant palaeoclimatic indicators for this period.

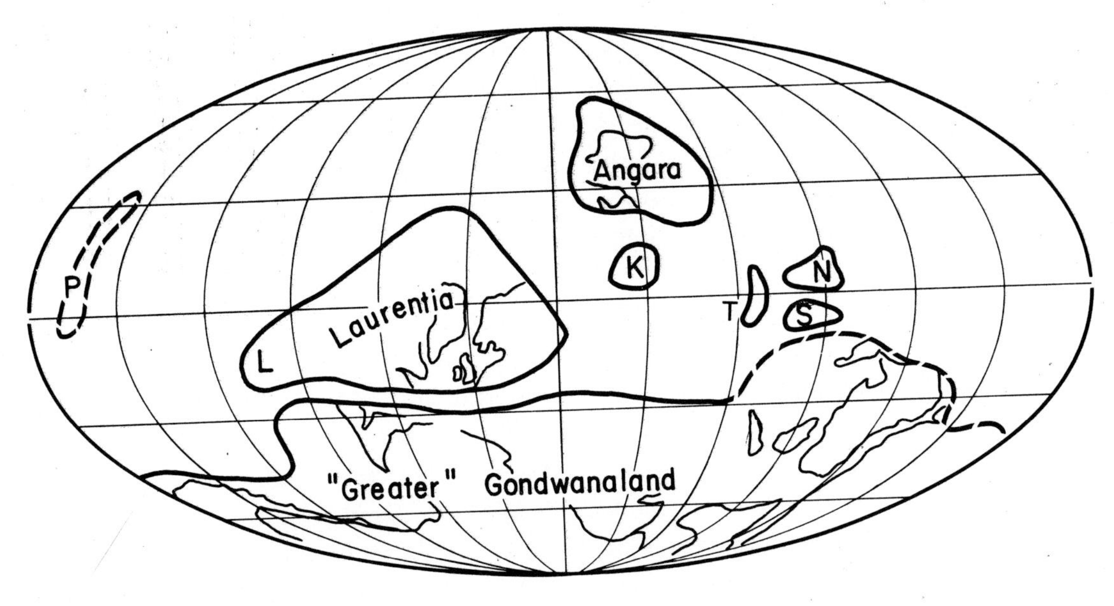

Fig. 2 A Lower Carboniferous reconstruction

See text for comments. Limited coastlines are outlined for location purposes only. It is also probable that the shapes of the continental units differed significantly from those shown, which are diagrammatic.

K=Kazakhstan; T=Tarim; N=North China; S=South China; P=Pacific.

OTHER APPLICATIONS

Palaeomagnetic techniques can be applied to a wide variety of geological problems, irrespective of the age of the rocks concerned.[1] Most such techniques are of general applicability, and not confined to Carboniferous times. For example, the detailed analysis of magnetic overprints associated with burial and the remanence acquired during uplift appear capable of placing strict controls on the timing and rate of vertical uplift of the crust. Such analyses are likely to be of particularly relevance to studying late and post-Variscan motions, in particular to an understanding of the origin and timing of the development of paralic basins following early Variscan orogenesis.

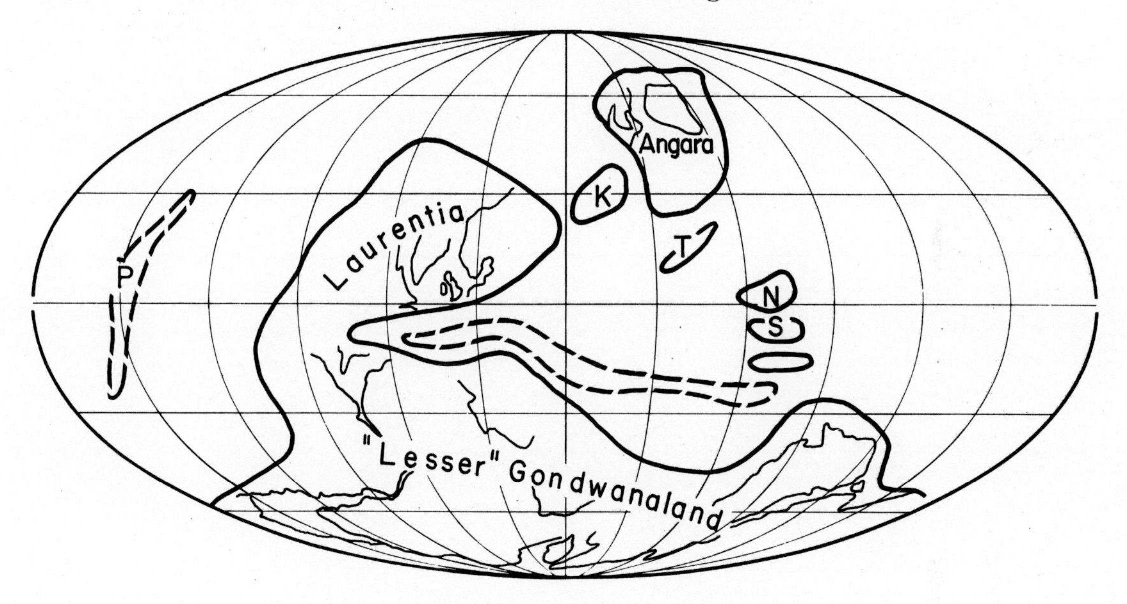

Fig. 3 An Upper Carboniferous reconstruction
(see text and the legend to Fig. 2)

REFERENCES

1. Tarling DH, Palaeomagnetism, Chapman and Hall, London, New York (1983), 379.
2. O'Reilly W, Rock and Mineral Magnetism, Blackie, Glasgow (1984), 220.
3. Collinson DW, Methods in Rock Magnetism and Palaeomagnetism, Chapman and Hall, London, New York (1983), 503.
4. Irving E and Parry LG, Geophys JR Astr Soc, 7 (1962), 395.
5. Palmer JA, Perry SPG and Tarling DH, Jour Geol Soc, London, 142 (1985), 945.
6. Perry SPG, Palaeomagnetic Studies of British Carbonate Sediments, Ph D Thesis, Univ Newcastle upon Tyne (1979).
7. Palmer J, Carboniferous Palaeomagnetism, Ph D Thesis, Univ Newcastle upon Tyne (1987).

8. Khramov AN, Paleomagnetology, Springer-Verlag (1987), 308.
9. Tarling DH, 10^{e} Cong Int Strat Geol Carb Madrid, 1983, C R 3 (1985), 153.
10. McElhinny MW, Embleton BJJ, Ma XH and Zhang ZK, Nature, 293 (1981), 212.
11. Cheng GL, Bai VH, and Li YA, Seimol Geol, 4 (1983), 12.
12. Lin Jinlu, Fuller M and Zhang Wenyou, Nature, 313 (1985), 444.
13. Bai Y, Chen G, Sun G, Sun Y, Li Y, Dong Y and Sun D, Tectonophys, 139 (1987), 145.
14. Lin Jinlu, Sci Geol Sinica, (2) (1987), 183.
15. Johnson GAL and Tarling DH, 10e Cong Int Strat Geol Carb Madrid, 1983, C R 3 (1985), 163.
16. Tarling DH, Gondwana and Tethys In: Audley-Charles MG and Hallam A (eds), Geol Soc London (1987) (in press).

XIe Congrès International de Stratigraphie et de Géologie du Carbonifère Beijing 1987,
Compte Rendu 1 (1991): 213-245

MAJOR SUBDIVISIONS OF THE CARBONIFEROUS SYSTEM

R. H. Wagner
(Jardin Botànico de Córdoba, Apartado 3048, E 14071 Cordoba, Spain; and ENCASUR, Pēnarroya-Pueblonuevo (Córdoba), Spain)

Cor F. Winkler Prins
(Rijksmuseum van Geologie en Mineralogie, Hooglandse Kerkgracht 17, 2312 HS Leiden, The Netherlands)

ABSTRACT

The existing chronostratigraphic subdivisions of the Carboniferous are discussed as well as their correlations and the position of major stratigraphic boundaries. The effects of climatic changes are considered for the mid-Carboniferous (Mississippian/Pennsylvanian) boundary and the Carboniferous/Permian boundary. Proposals are made for the approximate size of Carboniferous stages of worldwide use within the palaeoequatorial belt. The different faunal and floral developments of the Gondwana and Angara areas, due to their rather high palaeolatitudinal positions and consequently cool climates, make it advisable to create separate chronostratigraphic classifications for the Middle and Upper Carboniferous strata in these areas.

INTRODUCTION

The Carboniferous System combines two major intervals in the earth history, with different characteristics of sedimentation and floral and faunal distribution obtaining for these two intervals. Different criteria have to be applied for the subdivision of these two major intervals which seem to have corresponded to times with a very different world climate. The recent emphasis on finding a suitable mid-Carboniferous boundary[1] implies recognition of the fact that the change in physical (climatic) conditions in the world was reflected by changes in floral and faunal contents across this boundary and, also, in floral and faunal distribution patterns. This mid-Carboniferous boundary, separating two major stratigraphic units which Bouroz et al.[2] proposed to recognise as subsystems, might also be regarded as sufficiently important to warrant recognition as a systemic boundary like North American geologists have done for some time. These two major units have been called Mississippian and Pennsylvanian in North America, whereas similar though not entirely time-equivalent units in Western and Eastern Europe were called Lower and Upper Carboniferous (Dinantian and Silesian) and Lower and Middle + Upper Carboniferous.

The Mississippian corresponds to a time interval with a rather equitable climate and, indeed, only slight climatic differentiation. Sea level appears to have been generally high, as follows from the extensive areas of carbonate sedimentation on low-lying cratons like the Russian Platform, the North American Continent, and a large part of China.

Correlations of Mississippian strata are effectively worldwide as a result of the still rather slight differentiation of floras and faunas which allow the recognition of biozones across palaeolatitudinal belts. This is a continuation of the cosmopolitan faunas of Late Devonian times.[3] Stratigraphic correlation in the Mississippian is based almost entirely on marine faunas, but microfloras play an increasingly important role.

With the onset of the Pennsylvanian conditions changed. A marked climatic differentiation came into being. This is noted most clearly in the distribution of floras which are different in composition in various different latitudinal belts. Faunal differentiation also occurred, but the palaeolatitudinal effect of the faunal differentiation has probably been modified to some extent by ocean currents. A widespread regression took place after the Mississippian. This led to erosional breaks in the succession laid down on low-lying cratons and, most notably, to a change from carbonate to terrestrial sedimentation in some areas, e.g. large parts of the North American craton. Even where carbonate deposition continued, as on the Russian Platform, an erosional break is apparent. This lowering of sea level can only have had a climatic origin since this physical phenomenon is associated with a marked differentiation in floral composition in palaeolatitudinal belts.

The correlation of Pennsylvanian strata is therefore more difficult, since biozonations are necessarily linked to palaeolatitudinal belts. Not only are these zonations based on different floral and faunal associations, but the amount of diversification is strikingly different. The palaeoequatorial belt (approximately the tropics and subtropics) shows much more diversified floras and faunas than occur in the temperate and subarctic zones of the northern and southern hemispheres (Angara and Gondwana Realms). This means that the chronostratigraphic subdivision of Pennsylvanian strata in the palaeoequatorial belt can be much more detailed than may be possible for the higher latitude zones. Any attempt at a worldwide chronostratigraphic classification of Pennsylvanian strata must therefore take into account that only the broadest of subdivisions may be recognised all over the world and that a detailed subdivision may only be applied in the palaeoequatorial belt. In this respect the Pennsylvanian and, therefore, the greater part of the Carboniferous, is different to most systems.

SYSTEMIC BOUNDARIES

1. Devonian/Carboniferous boundary

The position of this boundary has been linked traditionally to certain faunal groups and, more recently, to microfloras. At the Second Carboniferous Congress, in Heerlen,

1935, the base of the *Gattendorfia* Ammonoid Zone was recommended as the base of the Carboniferous System.[4] No stratotype was proposed at the time since there was a general assumption that the time represented by the ammonoid zone and its associated fauna would serve as a stable synchronous unit for the basal Carboniferous. The more recent emphasis placed on stratotypes as a stable reference for stratigraphic correlation has made the *Gattendorfia* Zone less suitable. A stratotype containing goniatites is likely to include stratigraphic breaks and this implies problems with the boundaries of the requisite biozone in any particular locality.

The search for a suitable boundary stratotype has been actively taken into hand by the IUGS Working Group on the Devonian-Carboniferous Boundary. However, before selecting a stratotype, the working group has aimed at a biostratigraphic definition of the base of the Carboniferous System. A review of the biostratigraphic significance of the various groups of fossils involved has been published in 1984.[5] The same volume contains the recommendation that the base of the Carboniferous should be linked to the first appearance of *Siphonodella sulcata* (Huddle, 1934), which evolved from *Siphonodella praesulcata* Sandberg, 1972 of Devonian age. The incoming of this conodont immediately precedes the first occurrence of *Gattendorfia* in the Honnetal section of the Rhineland, West Germany, which served as the conceptual stratotype of the base of the Carboniferous when the *Gattendorfia* Zone was still regarded as the biostratigraphic marker.

Prior to the recommendation regarding *S. sulcata,* discussion had taken place on the merits of defining the base of the Carboniferous in accordance with the first occurrence of the foraminifer *Quasiendothyra kobeitusana* Rauser-Chernoussova, 1948. This reflected a historical controversy between ammonoid and foraminiferal workers. The latter carried most weight in the USSR where the base of the Carboniferous is generally placed at a lower horizon than that recommended at present.[6]

Paproth and Streel[5] also reported on the search for a boundary stratotype incorporating the biostratigraphic index, *Siphonodella sulcata,* and its probable ancestor, *S. praesulcata.* Three sections are mentioned as being capable of providing the boundary stratotype, viz. Hasselbach in the Rhenish Slate Mountains of West Germany, Berchogur in the Mugodzhar Mountains of Kazakhstan, and Muhua in South China. Another prospective stratotype, the Kiya section in Siberia, was mentioned but later withdrawn from the list of contenders.[5]

Generally speaking, the Devonian/Carboniferous boundary problem is a relatively minor one since one deals with cosmopolitan faunas in a fully marine environment, often in carbonate facies. The main problem seems to be making the right selection of a biostratigraphic marker within an evolutionary lineage that is uncontroversial. This may not be the case for *Siphonodella sulcata* which is not always easy to distinguish from its predecessor, *S. praesulcata.* The selection of the eventual stratotype should depend on the degree of exposure, accessibility, and the range of organisms present in addition to *S. praesulcata* and *S. sulcata.*

2. Carboniferous/Permian boundary

This boundary poses more major problems than the base of the Carboniferous.

These problems have not yet been tackled by a special working group of the IUGS Commission on Stratigraphy. Several different criteria play a role for the Carboniferous/Permian boundary. One of these is historical and relates to the evolutionary grade of Palaeotethyan marine organisms. Different positions for the actual boundary have been proposed in the area of warm-water faunas, depending on the kind of organisms and the inherent bias of specialist workers.[7] Another criterion concerns the climatic change operating at the end of the Pennsylvanian Ice Age. This produced an amelioration (warming) of the climate which was particularly noticeable in the high latitude areas of Gondwana and Angara, and a concomitant increase in aridity in the low-latitude palaeoequatorial belt area.

The warming effect produced an increased diversity of floral elements due to the influx of thermophile plants into the higher latitude areas. In the Gondwana Realm the Carboniferous/Permian boundary is usually drawn at the incoming of the glossopterids;[8] for the Angara Realm there is a more complex suite of floral elements which marks the climatic change.

The terrestrial facies of the low-latitude palaeoequatorial belt area have usually been correlated on the presence and relative increase of *Callipteris* and the conifers which means the presence of better drained soils in the area of sedimentation (and the relative absence of hygrophile plant assemblages). However, in certain parts of the palaeoequatorial belt the humid conditions of Pennsylvanian times continued (e. g. over most of China and, apparently, in the Middle East) and the *Callipteris/* Conifer Association is very rarely found.

The warm water marine facies and the terrestrial facies of the palaeoequatorial belt are bridged to some extent by microfloras which allow correlation. In this way the Russian Asselian is correlated to part of the Autunian of Western Europe. Since the middle Autunian microfloras (or new style lower Autunian of Bouroz and Doubinger[9]) show the increase of conifer-derived pollen grains which may be associated with the lessening of a humid climate in the palaeoequatorial belt, this apparently provides the means of linking a fairly generally accepted boundary on warm water faunas to the climatic change at the end of the Pennsylvanian. It is likely that this is the same change that produced the increased diversity of floral elements in the high latitude areas of Gondwana and Angara. This would appear to be a useful criterion for fixing the Carboniferous/Permian boundary at this level. Different biozones would operate in different parts of the world at this level but the climatic connotation conveyed by the floral remains would allow a broad correlation.

The Permian System was first described from the European part of the present USSR and its base was placed at the top of the *Schwagerina* Horizon. Murchison,[10] who introduced the system, drew the boundary between the Carboniferous and the Permian in the Kovrov area (Klyazma River) and in the Samara Bend in the South Urals. This area is still regarded as the conceptual stratotype, but the *Schwagerina* Horizon is presently considered to be basal Permian rather than Carboniferous, i.e. since Ruzhencev[11] erected the Asselian Stage. He correlated this stage with the *Schwagering* Horizon and with the Wolfcampian of North America. The base of the Asselian, which approximates to the base of the middle Autunian (or lower Autunian of Bouroz and Doubinger[9]), is the most generally accepted Carboniferous/Permian boundary in the USSR.[12] Its apparent correlation to the boundaries proposed on climatic change, as expressed by micro- and megafloral assemblages, makes it still the most attractive level at which the Carboniferous/Permian boundary may be drawn.

However, there are alternative boundaries as discussed by Rauser-Chernoussova and others.[7,12,13] These are based mainly on the distribution of fusulinid faunas. A boundary at a lower level than the Asselian is that below the *Daixina sokensis* Zone. At the base of this zone lies the first occurrence of *Pseudofusulina*, which is the direct ancestor to the Schwagerinidae, a typical Permian group of fusulinid foraminifers. This is why certain authors prefer the base of the *Daixina sokensis* Zone to coincide with the base of the Permian. An even lower boundary, at the base of the *Eotriticites montiparus* Zone, was advocated by Zhang[14] on the basis of his interpretation of the evolution of fusulinid wall structure.

A position higher than the base of the Asselian has been proposed by Barkhatova,[15] who analysed the faunal remains in different parts of the *Schwagerina* Horizon (fusulinids and ammonoids in particular) and found that the faunas changed most significantly at the base of the upper zone of the Asselian, i.e. the Nenetsky Horizon. Another boundary was taken at the top of the Asselian[16,17] since the Schwagerinidae characterise this unit as well as the underlying Gzhelian and appear already in the Kasimovian.[15] The top of the Asselian is also the base of the Sakmarian which is the historical base of the Permian System.

An even higher position of the base of the Permian is advocated by a majority of the Chinese workers who include most of the Sakmarian with the Upper Carboniferous.[15] The reasons for placing the Carboniferous/Permian boundary at this high level are not very clear but they seem to have to do with regional mapping.[18] Recently, however, Chinese fusulinid workers have proposed a lower boundary, close to the base of the Asselian.[7,19]

In the terrestrial development of Upper Carboniferous and Lower Permian in Western Europe the Autunian is generally regarded as Permian and the underlying Stephanian (Series) as the highest Carboniferous. The Autunian as represented in the Autun Basin, in the northeastern part of the Massif Central, France, was originally subdivided into lower, middle and upper Autunian, with the Grand Molloy and Igornay beds being assigned to lower Autunian and the Nuse Beds to the middle Autunian. A correlation with the type Stephanian (upper part) in the St Etienne region, further south in the Massif Central, has suggested that the lower Autunian overlaps part of the upper Stephanian. The area of overlap contains already *Callipteris* and several species of the conifer *Lebachia* (*Walchia*), and this has been called the Stephanian D by Bouroz and Doubinger,[9] a unit of dubious validity since it was taken out of the previously defined Stephanian C. Wagner[20] has preferred to recognise this unit as a biozone, the *Callipteris conferta* Zone. Since Bouroz and Doubinger recognised the Grand Molloy and Igornay beds as Stephanian, the base of the Autunian in its stratotype had to be lifted to the level of the Muse Beds, this being coincident with the *Potonieisporites* Zone which apparently marks the increase in conifer pollen at the climatic change at the end of the Pennsylvanian Ice Age. The *Potonieisporites* Zone also seems to coincide with the lower part of the Asselian. However, the Asselian mentioned by Bouroz and Doubinger[9] does not correspond to the stratotype of this basal Permian stage but to beds of equivalent age in the Donets Basin.

The Carboniferous/Permian boundary in the high latitude Gondwana and Angara areas may also be interpreted in the light of a climatic change. A straightforward correlation between palynozones in these high latitude areas and in the palaeoequatorial belt region is not possible since the different proportions between the various groups of plants will relate to environmental conditions in which the climatic factor plays an important role. Generally speaking, the improvement of climatic conditions in the high latitude areas will have led to the introduction of thermophile plants and an increase in floral diversity, if compared with the Pennsylvanian floras which lived under harsher conditions. This development at the end of the Pennsylvanian probably occurred at virtually the same time as the general introduction of more arid conditions in parts of the palaeoequatorial belt, which led to specialised plant assembalges in these drier tracts (it should be noted that not the entire palaeoequatorial area became drier and that humid condition persisted in Permian times over large regions, such as China, Southeast Asia, the Middle East, and Venezuela to the southern United States).

The notable increase in the proportion of *Potonieisporites* (c. 80%) in the assemblage at the base of the Autunian sensu Bouroz and Doubinger,[9] reflecting the rather sudden spreading of more arid conditions in parts of the palaeoequatorial belt (with the so-called Atlantic or Euramerican floras), can probably be correlated with the increase in floral diversity recorded at the end of the Pennsylvanian in the Gondwana area. This is the change-over from the *Potonieisporites* Zone to the *Cristatisporites* Zone as described by Archangelsky & and Marques Toigo.[21] The *Potonieisporites* Zone of the Gondwana area would therefore be older than the *Potonieisporites* Zone of the more arid parts of the palaeoequatorial belt. Both *Potonieisporites* zones would reflect somewhat adverse conditions occurring at different times in two different major areas.

In the Angara area, in the temperate belt on the northern hemisphere of Carboniferous and Permian times, the Carboniferous/Permian boundary is traced on the transition from a predominantly cordaitean assemblage with a minor contribution of pteridosperms, ferns and sphenophytes, to a gymnosperm assemblage (of the lower and upper Balakhonsk, respectively). The climatic change does not seem to be quite as clearly marked in this area as in the Gondwana Realm.[22]

STRATIGRAPHIC CLASSIFICATION OF THE CARBONIFEROUS

1. General remarks

Climatic differentiation which slowly increased during the Early Carboniferous (Mississippian) and which was notably present from the Middle Carboniferous (early Pennsylvanian) onwards, has led to serious difficulties for the recognition of chronostratigraphic units worldwide. In fact, only the most major chronostratigraphic units (subsystems, series) may have a chance of being recognised all over the world, i.e. at different palaeolatitudes. This makes the Carboniferous unusual among systems. In practice, the smaller chronostratigraphic units (stages, substages) have only been applied successfully in the palaeoequatorial belt (approximately the tropics and subtropics), which contains the most varied floras and faunas. Correlations with the (cold) temperate and subarctic southern and northern hemisphere areas (Gondwana and Angara Realms) are recognised as being only approximate for Middle and Upper Carboniferous strata, since faunal and floral assemblages are generally different and much less diversified. The existing difficulties are not always fully apparent since attempts have been made in Australia and South America to apply West European (palaeoequatorial) chronostratigraphic units, and in the Angara area to distinguish units based on the stratigraphic development in the Russian Platform region. These attempts, which have been more or less successful with regard to the Lower Carboniferous (Mississippian), do not seem to have led to convincing results for the Middle and Upper Carboniferous (Pennsylvanian). It is therefore apparent that regional chronostratigraphic classifications will have to be developed for Middle and Upper Carboniferous strata in the Gondwana and Angara Realms. Due to the relatively poor diversification of floras and faunas in these southern and northern hemisphere areas, their chronostratigraphic units are bound to be larger and fewer than those of the palaeoequatorial belt. Partial recognition of the need for separate stratigraphic classifications in the Gondwana and Angara Realms is apparent by the existence of regional biozonations, but no chronostratigraphic schemes for these realms have been developed as yet.

The classical areas for Carboniferous stratigraphy lie in the palaeoequatorial belt where well diversified floras and faunas exist. The lack of a unified stratigraphic scheme in the palaeoequatorial belt is mainly due to historical reasons and, to a lesser extent, to different basinal histories leading to different facies successions in the different regions where the chronostratigraphic classifications were developed. Two rival schemes have been developed in the USSR and in Western Europe, whereas the North Americans developed different classifications based on marine and generally non-marine successions in the Mid-Continent, Illinois Basin and Appalachian Basin. Eventually, the United States Geological Survey adopted a classification into regional series which relies primarily on marine fossils for correlation.

The regional chronostratigraphic units in the different areas of the palaeoequatorial belt are not of equivalent size due to historical reasons and the different kinds of fossils regarded traditionally as the most important in these different area. It is apparent that the stages in the USSR scheme are larger than they need to be for use as effective chronostratigraphic units within the palaeoequatorial belt. On the other hand, the much smaller stages of the West European scheme may not always serve as effective chronostratigraphic units to be recognised throughout the palaeoequatorial belt. An integrated chronostratigraphic classification capable of being applied throughout the palaeoequatorial belt should adopt a compromise solution with stages of approximately half the size of those recognised presently in the USSR.

Bouroz et al.[2] made a first attempt to propose an integrated classification, dividing the Carboniferous System into two subsystems named Mississippian and Pennsylvanian, and subdividing the latter into two series. This seemed to marry successfully the American usage of recognising two systems, Mississippian and Pennsylvanian, with the threefold division into Lower, Middle an Upper Series customary in the USSR. There is a general agreement that the Lower/Middle Carboniferous boundary (Mississippian/Pennsylvanian boundary) is more easily defined in biostratigraphic terms than the Middle/Upper Carboniferous boundary. The exact position of this mid-Carboniferous boundary has been defined at the meeting of the IUGS Subcommission on Carboniferous Stratigraphy in Madrid, 1983.[23,24] This boundary was linked to the evolutionary appearance of the conodont species *Declinognathodus noduliferus* (Ellison et Graves, 1941) and is low in the lower Namurian Series of the West European chronostratigraphic scheme. It lies below the currently accepted Lower/Middle Carboniferous boundary of the USSR, whereas the Mississippian/Pennsylvanian boundary of North America does not have a formally accepted position and therefore seems adaptable to the decision taken in Madrid. There is no accepted stratotype for the Mississippian/Pennsylvanian boundary (mid-Carboniferous boundary) either in America or elsewhere, and this is a major challenge currently taken up by the mid-Carboniferous boundary committee of IUGS-SCCS.

Bouroz et al.[2] highlighted the world-wide palaeogeographic changes and marked climatic differentiation which took place as a result of the Carboniferous (Pennsylvanian) Ice Age. This major climatic event is bound to have affected the composition of floras and faunas. Widespread regression affects the distribution of shallow water faunas and climatic differentiation will have provoked a marked palaeolatitudinal zonation of land floras (as is clearly apparent from the fossil record). It is likely that the climatic change will have provoked an instant response with regard to the composition of continental floras, whereas the marine faunas may have shown a somewhat retarded effect. The first floral assemblages of Pennsylvanian aspect appear at the level of the highest Arnsbergian (lower Namurian, E_2) in Western Europe,[25,26] but the record is generally poor at this level. Faunas show generally changes from Chokierian (*Homoceras* Zone) onwards.[27]

There is no world-wide event justifying the selection of a major boundary at the level of the base of the Upper Carboniferous in the Russian sense or the Stephanian of Western Europe. Evidently, a boundary at this level cannot be regarded as being of the same rank as that between Mississippian/Pennsylvanian and Lower/Middle Carboniferous. The base of the West European Namurian Series is essentially a facies change in NW Europe where limestone deposition changed into terrestrial facies and has no world-wide significance (as is generally acknowledged in the recent literature).

Bouroz et al.[2] consequently proposed the Mississippian/Pennsylvanian boundary as one between subsystems, whereas the Middle/Upper Carboniferous boundary (in the

Russian sense) was regarded as a boundary between series. Stage boundaries were not discussed in that paper which merely aspired to provide a general framework for discussion.

In the following chapters the three rival schemes (West European, USSR, North American) will be discussed, as well as the other major subdivisions.

2. Western Europe

This classification[28] is largely influenced by the sedimentary history of the Northwest European paralic coal belt, a large shelf area extending from the British Isles across the North Sea to the Low Countries, Germany and Poland. Within this relatively stable area with a moderate amount of subsidence, there are local fracture zones limiting downwarps and relative uplifts (arches) which give local variations between basins and blocks. The subsiding shelf region bordered north-northwestwards onto a stable continental area (North Atlantean Continent) which at least partly coincided with Scandinavia and which may have been a carbonate platform with apparent links with the Russian Platform.[29,30] South-southeastwards the shelf region bordered onto the mobile Rhenish-Hercynian Zone which acted as an active hinterland and which imposed a marked polarity on tectonic structure and successive depocentres which migrated NNW-wards (Teichmüller[31] and others). Westphalian sediments, which are markedly widespread and quite uniformly developed, show a general thinning and fining NNW-wards. The widespread marine transgressions of Westphalian times show the open sea to have been to the southwest (faunas from the southwestern part of the British Isles are most varied and show affinities with the Cantabrian Mountains (NW Spain), the western termination of the Palaeotethys).[32] Local variations in thickness and facies indicate that the subsiding shelf region was almost featureless in Westphalian times.[33]

Although there were local turbidite basins in SW England and in the Sauerland of Germany in the Early Carboniferous, the classical Lower Carboniferous (Dinantian) facies is the Carboniferous Limestone (Mountain Limestone), a carbonate development linked to the Wales-Brabant Massif, a local arch (or series of arches) near the edge of the subsiding shelf region, and to local uplifts (blocks) further north in the general shelf area. Terrigenous facies became widespread in Namurian times (after tectonic movements of the so-called Sudetic Phase) and developed from almost wholly marine to alternating marine and terrestrial, forming a gradual transition to the Westphalian which is mainly in terrestrial facies with occasional marine transgressive intervals of a eustatic nature.

Local orogenic movements seem to have taken place before Westphalian D and it appears that Westphalian D (and lower Cantabrian) basins were not quite as widespread but more locally developed than the area of sedimentation of Westphalian A, B and C. In later Stephanian times the Northwest European paralic coal belt was extensively modified by folding and thrusting in the internal part and essentially vertical tectonics in the main part of the mobile shelf region. The Stephanian is generally absent (published records of Stephanian strata are most often questionable) and subsequent lower Rotliegend (highest Stephanian to lowermost Permian) strata show a distribution which is markedly different to that of the underlying Westphalian. A large stratigraphic gap (equivalent to at least ten million years) is accompanied by fundamental changes in palaeogeography.

The stratigraphic classification elaborated by the Heerlen Congress is closely linked to the sedimentary history of the subsiding shelf region. The subsystemic boundary between Dinantian and Silesian[34] reflects the change from carbonate deposition to more widespread terrigenous facies after local movements of the Sudetic Phase produced palaeogeographic changes. Although this boundary has later been defined in biostratigraphic terms at the incoming of *Cravenoceras leion* Bisat, 1930, there is no major renovation of faunas and floras at this level but merely a regional facies change which led to a replacement of coral-brachiopod faunas of the carbonate shelf by bivalve-goniatite assemblages. The base of the Westphalian Series is defined by a major marine transgression, but the original justification for distinguishing Namurian from Westphalian lies in the general absence of workable coal seams in the former. Again there is no apparent renovation of faunas and floras at this boundary.

Goniatites have been used as the preferred fossil group in the terrigenous facies of the Dinantian and in the Namurian where successive marine transgressions brought in successive assemblages.[35] Although evolutionary changes have been used for a goniatite biostratigraphy, the main emphasis has been on successive faunas in goniatite bands formed at the height of transgressions. This use of the goniatite faunas is most apparent within the Westphalian Series where the base of the Langsettian (Westphalian A), Duckmantian (Westphalian B) and Bolsovian (Westphalian C) Stages are determined by marine bands which are characterised by different goniatite assemblages. In the latter case the stratigraphic ranges of the goniatites found in these marine bands are not well known. The assemblages are poorly diversified if compared with the faunas occurring at comparable levels elsewhere (e.g. in North America and USSR) and correlation relies on the presence of the eustatic marine transgressive horizons as such. The widespread nature of these marine bands allows them to be used as virtually synchronous marker bands. It is clear that the relative stability of the shelf area producing an almost featureless plain, allowed these eustatic marine transgressions to sweep across a vast area and to produce lithological markers which can be identified from the British Isles across the Low Countries into Germany. Although it is likely that these transgressions correspond to interglacials and thus may be of world-wide significance, they are difficult to fit into a biostratigraphy and cannot be identified easily where a continuous marine succession exists (and where the stratigraphic ranges of individual biota are used for correlation). The erroneous results of the attempts by Ross[36] to use transgressions for the correlation of the major chronostratigraphic schemes of the Pennsylvanian may serve as an example.

In the upper part of the Westphalian no marine bands are found and purely biostratigraphic criteria are used to identify the base of the Westphalian D Stage. This stage, although represented in certain parts of the NW European paralic coal belt (South Wales, Bristol/Somerset, Oxfordshire, Osnabruck area, etc.), has been defined originally in the Saar-Lorraine Basin, well inside the Mid-European hinterland (Saxo-Thuringian Zone).[37]

Conodonts have come to the aid of the goniatite zonation, providing a more commonly present biostratigraphic tool of high precision. Miospores range across marine and terrestrial facies and have provided an effective biozonation[38,39] up to and including the Westphalian D, and plant megafossils have also proved to be effective in concurrent-range zones,[20] particularly in Westphalian strata. Non-marine bivalves have also been used in the Westphalian[40] where sedimentary conditions allowed the preservation of large assemblages at repeated intervals (Fig. 1).

The Dinantian carbonates were subdivided into Tournaisian and Visean Stages which were later regarded as series. Stages have been recognised in the British Isles, Belgium and West Germany, with increasing emphasis on the British scheme (Fig. 1), but no formal recognition has been given to these stages in the West European chronostratigraphy. The Namurian, also originally regarded as a stage, is presently recognised as a series with seven stages which are coincident with goniatite zones E_1, E_2, H_1, H_2, R_1, R_2, and G_1 (Fig. 1). Biostratigraphers working with terrestrial strata containing floral remains have often used the informal units A. B and C, and this usage has persisted where the formal stages proved difficult to recognise. The Westphalian, also used originally as a stage with substages A, B, C, and later, also D, is a series since Krefeld, 1971.[45] The substages becoming stages have needed formal names, and Langsettian, Duckmantian and Bolsovian have been proposed by Owens et al.[43] for the A, B and C units. Boundary stratotypes for all the stages in the Namurian have been recognised in Great Britain and Ireland[46,47] (Arnsbergian and Chokierian, despite being names derived from German and Belgian localities, have their stratotypes in the British Isles). The boundary stratotypes for the three Westphalian stages mentioned above are also in Britain where permanent exposures exist.[46,48]

The Westphalian D, introduced by Bertrand[49] for the upper part of the productive Westphalian in Saar-Lorraine, has been based on entirely terrestrial deposits with flora. There is no formally recognised stratotype for this unit, the base of which has been linked to the incoming of certain plant fossils (e.g. *Neuropteris ovata* Hoffmann).

The Stephanian Series, at the top of the Carboniferous, has been based on the limnic basins of the Massif Central in France, with three divisions, A, B and C being recognised.[50] A Stephanian D unit was carved out of Stephanian C and lower Autunian by Doubinger,[51] but failed to gain formal acceptance. Plant fossils provided the means for biostratigraphic correlation, whereas volcanic ash bands (cineritic tonsteins[52-53]) served as lithological markers within the general region, linking the individual basins.

It has been long recognised that a time gap existed between Westphalian D and Stephanian A, but the amount of time missing was not apparent until an uninterrupted succession bridging the gap was recorded from the Cantabrian Mountains in northern Spain. This led to the introduction of the Cantabrian Stage as the basal unit of the Stephanian Series.[44]

The limited nature of biological elements present in the Stephanian deposits of south-central France (fossil flora, fish remains, bivalves, and other non-marine elements) has made the limnic basins of the Massif Central unsuitable for stratotypic purposes. More suitable Stephanian successions are developed in the Cantabrian Mountains where marine strata alternate with terrestrial deposits. The official stratotypes of the two lower

Fig. 1 Chronostratigraphic classification and biozonations for Western Europe[20,28,38-44]

	Chronostratigraphic Units		Belg. lith. units	Ammon. Zones	Conodont Zones	Foraminiferal Zones	Miospore Zones	Megafloral Zones	Non-marine Bivalve Zones
L. P. Aut.								Callipteris conferta	
SILESIAN	Stephanian	C (no formal name)					P. novicus-bhardwaji	Sphenophyllum angustifolium	
		B (no formal name)					C. major	Alethopteris zeilleri	
		(A) Barruelian					A. splendidus	Lobatopteris lamuriana	
		Cantabrian					T. obscura-thiesseni	Odontopteris cantabrica	
	Westphalian	D (no formal name)						L. vestita Linopteris obliqua Paripteris linguaefolia	A. prolifera A. tenuis A. phillipsii u. similis-pulchra l. similis-pulchra
		(C) Bolsovian					T. securis-laevigata M. nobilis	Lo. rugosa- A. urophylla	A. modiolaris
		(B) Duckmantian		'A'			F. junior	L. hoeninghausi-	C. communis
		(A) Langsettian		G2	Id. sulcatus parvus		R. aligerens C. saturni T. sinani	N. schlehani	C. lenisulcata
	Namurian	Yeadonian		G1	S. nodosum I. primulus		R. fulva R. reticulatus	N. larischi- Pecopteris aspera	
		Marsdenian		R2	Id. corrugatus		C. kosankei G. vario-reticulatus		
		Kinderscoutian		R1	Id. sulcatus		L. subtriquetra	L. larischi	
		Alportian		H2	Id. noduliferus		K. ornatus		
		Chokierian		H1	S. lateralis		S. triangulus	L. berm. - L. stangeri	
		Arnsbergian		E2	G. bilin. bollandensis {upper, lower}		R. knoxi B. nitidus		
		Pendleian		E1	{Kladognathus {G. girtyi simplex	Eosigmoilina	K. carnosus	L. bermudensiformis – N. antecedens	
DINANTIAN	Visean	Brigantian	V3c	Goβ-γ	P. nodosus		T. vetustus R. fracta R. nigra		
		Asbian	V3b	Goα	G. bilineatus	Asterodiscus	T. marginatus P. tessellatus		
		Holkerian	V3a V2b	Peδ		Koskinotextularia- Qu. ? nibelis	S. campyloptera K. triradiatus K. stephonephorus	Triphyllopteris	
		Arundian	V2a V1b		P. commutatus	Eoparastaffella	L. pusilla		
		Chadian	V1a		G. homopunctatus				
	Tournaisian	Ivorian	Tn3	Peγ Peβ	Sc. anchoralis P. communis carina	Tetrataxis Tournayella	S. claviger A. macra S. pretiosus R. clavata		
		Hastarian	Tn2 Tn1b	Peα Ga	Siphonodella crenulata S. duplicata	Chernyshinella	S. balteatus R. polyptycha K. hibernicus U. distinctus		
					S. sulcata		V. verrucosus R. incohatus	Adiantites	
U. D. Fam.		Strunian	Tn1a Fa2d	Wo	S. praesulcata	Quasiendothyra	S. lepidophytus V. nitidus		

Stephanian stages, the Cantabrian and Barruelian (the latter proposed in Madrid, 1983, to replace the informally named Stephanian A) are in the Cantabrian Mountains,[44] and research is being done to examine the suitability of this area for the recognition of stratotypes of the two upper Stephanian Stages, still informally named Stephanian B and C. The presence of fusulinid foraminifers, brachiopods and other marine faunas throughout most of the Spanish Stephanian makes it possible to effect correlations with the highest Moscovian, Kasimovian and Gzhelian of the USSR. The Carnic Alps at the Austrian/Italian border, another area with plant fossils and marine faunas including fusulinids,[54] seems to be too complicated tectonically to be very useful as a stratotype area.

Traditionally, the Stephanian in south-central France is followed by the Autunian, a non-marine unit of largely Early Permian age. Bouroz & and Doubinger[9] have proposed a new definition of the basal Autunian which allows correlation with the Asselian of the USSR. The biostratigraphic recognition of the basal Autunian has been linked to a sudden increase of the conifer pollen *Potonieisporites,* and it is likely that this biological effect is due to climatic changes related to the waning of the Pennsylvanian Ice Age. As such, this boundary would have considerable significance for world-wide chronostratigraphy.

In summary, the West European chronostratigraphic classification is linked to a large extent to the sedimentary development in different basins (NW European paralic coal belt, Saar-Lorraine, Massif Central in France, Cantabrian Mountains), lacks the variety of facies and biological contents required for a truly international scheme, and shows an uneven reliance on biostratigraphic criteria. The Spanish area is probably the most satisfactory for international correlation in the Westphalian and Stephanian, whilst the British Isles serve reasonably well for the Namurian interval and, perhaps, also the Dinantian.

3. USSR

In the USSR the West European classification is adopted for the Tournaisian and Visean (here considered as stages, whereas they are regarded as series in Western Europe), with the same emphasis on successive foraminiferal faunas. Indeed, work on these faunas was more advanced in the USSR and the results obtained in this area have been transferred to Western Europe where the application of foraminiferal studies has proved equally satisfactory.[41] A major difference is the position of the Devonian/Carboniferous boundary which is placed at the base of the *Quasiendothyra kobeitusana* Zone in the USSR (i.e. lower than in Western Europe).

The West European Namurian was used in a restricted sense, i.e. up to the base of the Bashkirian which includes the upper and middle Namurian (depending on the position of the base of the Bashkirian which was lowered in 1974 to include the middle Namurian *Reticuloceras* Zone). The (lower) Namurian was replaced by the Serpukhovian, the top stage of the Lower Carboniferous, the latter being more or less equivalent to the North American Mississippian.

The Middle Carboniferous is constituted by the Bashkirian and Moscovian Stages, the former being based on the South Urals and the latter on the central part of the Rus-

		Unified scheme Russian Platform	Donbass	Urals	Foraminiferal Zones
L. P.	ASS.	Sokoljegorsky			Schwagerina vulgaris - Sch. fusiformis
UPPER CARB.	GZHELIAN	Noginsky			Daixina sokensis
		Pavlovo - Posadsky			Jigulites jigulensis
		Amerevsky			Triticites rossicus - T. stuckenbergi
		Rusavkinsky			
	KASIMOVIAN	Yauzsky	C_3^c		T. acutus - T. quasiarcticus
		Dorogomilovsky			
		Khamovnichesky	C_3^b		Eotriticitis montiparus
		Krevyakinsky	C_3^a		Protriticites pseudomontiparus - Obsoletes obsoletus
MIDDLE CARBONIFEROUS	MOSCOVIAN	Myachkovsky	C_2^m d-e	Lazarevsky	Fusulinella bocki - Pulchrella eopulchra - Fusulina cylindrica
		Podolsky	C_2^m c	Kumyshsky	F. colaniae - F. vozhgalensis - Beedeina kamensis
		Kashirsky	C_2^m b	Kremensky	F. subpulchra - Hemifusulina kashirica
		Tsninsky		Elovsky	Aljutovella priscoidea - A. znensis - H. volgensis
		Vereisky	C_2^m a		A. aljutovica - Schubertella pauciseptata
	BASHKIRIAN	Melekessky	C_2^b e	Asatausky	Verella spicata - A. tikhonovichi
		Cheremshansky	C_2^b b-d	Tashastinsky	Ozawainella pararhomboidalis - Profusulinella primitiva
		Prikamsky	C_2^b a	Askynbashsky	Pseudostaffella praegorskyi - Pr. staffellaeformis
		Severokeltmensky	C_1^n e	Akavassky	Ps. antiqua
		Krasnopolyansky	C_1^n b-d	Syuransky	Eostaffella pseudostruvei - E. postmosquensis
LOWER CARBONIFEROUS	SERPUKHOV.	Voznesensky	C_1^s g	Bogdanovsky	Plectostaffella bogdanovkensis
		Zapaltyubinsky	C_1^s f C_1^s e	Brazhkinsky	Eostaffellina protvae Eosigmoilina explicata Monotaxinoides subplana
		Protvinsky	C_1^s d		
		Steshevsky	C_1^s c C_1^s b	Kosogorsky	Pseudoendothyra globosa Neoarchaediscus parvus
		Tarussky	C_1^s a		
	VISEAN	Venevsky	C_1^v g	Kurmakovsky	Endothyranopsis crassa Archaediscus gigas
		Mikhailovsky	$C_1^v f_2$ $C_1^v f_1$	Ladeininsky	
		Aleksinsky	$C_1^v e_2$	Gubashkinsky	
		Tulsky	$C_1^v e_1$	Kurtymsky	Endothyranopsis compressa Propermodiscus krestovnikovi
		Bobrikovsky	$C_1^v d_2$	Shishikhinsky	Uralodiscus rotundus Ammoarchaediscus primaevus
		Radaevsky	$C_1^v d_1$ C_1^v b-c	Kluchevsky	Eoparastaffella simplex Eoendothyranopsis
		Elkhovsky	C_1^v a	Kosvinsky	Endothyra elegia Eotextularia diversa
	TOURNAISIAN	Kizelovsky	C_1^t d	Kizelovsky	Spinoendothyra costifera Tuberoendothyra tuberculata
		Cherepetsky	C_1^t c	Kosorechinsky	Chernyshinella glomiformis
		Upinsky	C_1^t b	Kalapovsky	Eochernyshinella
		Malevsky		Lytvinsky	Bisphaera malevkensis Earlandia minima
		Zavolzhsky	C_1^t a		Quasiendothyra kobeitusana Q. communis
U. D.					

Series	Stage	Conodont Zones	Ammonoid Zones	Miospore Zones	Megafloral Zones
L. P.	ASS.			PT	
UPPER CARB.	GZHELIAN		Shumardites-Vidrioceras (?)	M	
			Dunbarites-Parashumardites	FL-TT	
	KASIMOVIAN	Streptognathodus gracilis-S. elegantulus			
		S. oppletus-S. excelsus		S-VQ	Odontopteris subcrenulata
MIDDLE CARBONIFEROUS	MOSCOVIAN	S. cancellosus-Neognathodus roundyi	Pseudoparalegoceras-Wellerites	C-VL TS-KH	Neuropteris scheuchzeri
		N. medexultimus-id. podolskensis			Linopteris neuropteroides-Dicksonites pluecken eti
		N. medadultimus N. bothrops	Paralegoceras-Eowellerites		Sigillaria transversalis Calamites carinatus
		N. bothrops-I. obliquus		WI-AG	Paripteris gigantea Annularia stellata
		I. fossatus	Diaboloceras-Winslowoceras		Alethopteris decurrens-Neuropteris rarinervis
	BASHKIRIAN	Idiognathoides incurvus	Diaboloceras-Axinolobus	EG-RA	Eusphenopteris trifoliolata-Sphenophyllum majus
			Branneroceras-Gastrioceras	RA-S	Ly. hoeninghausii-Eu. nummularia
		Id. sulcatus	Bilinguites-Cancelloceras	CC-MP	Neuralethopteris-Ly. hoeninghausii
			Reticuloceras Bashkortoceras	AP RM	Neuralethopteris-Karinopteris acuta Mesocalamites-Cordaites
LOWER CARBONIFEROUS	SERPUKHOV.	G. bilin. bollandensis Gnathodus girtyi	Homoceras-Hudsonoceras	CO-RA	
			Fayettevillea-Delepinoceras		
			Uralopronorites Cravenoceras	KD-GS P-T	
	VISEAN		Hypergoniatites-Ferganoceras	SR-PD	
				RF-LB	
			Beyrichoceras-Goniatites	LP-MB RE-LA	
		Scaliognathus anchoralis	Merocanites-Ammonellipsites	LP-EC	
	TOURNAISIAN	Dollymae bouckaerti			
		Si. crenulata	Protocanites-Pericyclus	RS-LM	
		Si quadruplicata			
		Si. sandbergi		AU	
		Siphonodella duplicata	Protocanites-Gattendorfia	HF	
		Si. sulcata	Wocklumeria	SL	
		Si praesulcata Bispathodus costatus			Lepidodendropsis
U. D.					

Fig. 2 Chronostratigraphic classifications and biozonations for the European part of the USSR[56,58-62]

sian Platform. These two units are roughly equivalent to the middle/upper Namurian and Westphalian of Western Europe, but the position of the Bashkirian/Moscovian boundary in West European terms is still subject to discussion.[55] The Upper Carboniferous (generally but not exactly equivalent to the West European Stephanian Series) is subdivided into the Kasimovian and Gzhelian Stages. Both are based on successions in the Moscow Syneclise (central Russian Platform).

The chronostratigraphic classification developed in European Russia is applied to the Asian part of the USSR, albeit with certain qualifications in the Angara Realm.

Local subdivisions are horizons and superhorizons which are rock units characterised by fossil assemblages. Suites are local lithostratigraphic units. Although Bashkirian, Moscovian, Kasimovian, and Gzhelian are based on rather thin marine (predominantly carbonate) successions of the Russian Platform, there is a thicker (and potentially more complete) development of strata of the same ages in the Donets Basin, south of the Russian Platform. Carbonate platform successions normally include stratigraphic gaps of different magnitudes, and with increasing resolution of the palaeontological method these gaps become apparent. They also show a restricted range of facies, emphasizing shallow marine environments. The more continuous development of strata in the Donets Basin, showing a more varied range of facies which include terrestrial environments, has tended to make the Donets succession a reference section for correlations outside the USSR.[56,57] Within the Donbass the subdivison into suites is related to calcareous intervals numbered alphabetically with subunits 1,2,3, etc. Lower case letters are used to identify coals witin the suites. For correlations within the USSR foraminifers (fusulinids above all) have been used preferentially, although brachiopods have also been used traditionally, and there has been increasing emphasis in recent years on the goniatites, conodonts and spores. Plant megafossils have been used in the Donets Basin, Caucasus and in Angaraland. Fig. 2 shows various subdivisions used in different parts of the USSR including biozones.

Gaps in the stratigraphic succession of the Moscow Syneclise include a major break at the Lower/Middle Carboniferous boundary where the highest Serpukhovian is missing. This gap may be related to the widespread regression which took place at the onset of the Pennsylvanian Ice Age. Another gap of major proportions exists at the Bashkirian/Moscovian boundary where lower Moscovian oversteps Serpukhovian and even Visean in the eastern part of the syneclise. A stratigraphic gap within the Moscovian succession has recently been recorded and filled by the Tsninsky Horizon which has been intercalated between the Vereisky and Kashirsky Horizons of the lower Moscovian.[62,63]

Correlation problems are partly due to lack of continuity in the succession in a stratotype and partly to limitations in fossil content. Indirect correlations via the Donets Basin have led to problems with the positioning of the Bashkirian/Mosciovian boundary in the West European chronostratigraphic succession.[55] Solovieva et al.[62] mention that the terrigenous strata underlying the fusulinid-bearing carbonate sequence of the type Vereisky (basal Moscovian) have yielded spores similar to those of Assemblage II of Pegusheva[64] from a locality in the southwestern part of the Russian Plate where spore-bearing strata alternate with limestone layers containing Vereisky foraminifers. The carbonates of the upper part of the type Vereisky have not only yielded foraminifers, but also conodonts which allow correlation with the K_1-K_2 interval of the Donets Basin.[62] It is noted that this interval was attributed originally to the highest Bashkirian.[56] Wagner and Bowman[55] have presented evidence for a much lower position of the Bashkirian/Moscovian boundary in NW Spain than has been suggested by the existing correlations in the Donets Basin, where suggestions range from the base of Westphalian C to Westphalian B. The Spanish information suggests a position within Westphalian A. The stratigraphic gap at the base of the Vereisky in its stratotype prevents a full succession of fusulinid faunas from being recorded in that area. This gap is presently being bridged in the Cantabrian Mountains (NW Spain) where transitional faunas between Bashkirian and Moscovian occur.[65]

Another stratigraphic gap is associated with the base of the Kasimovian Stage (Upper Carboniferous) in the stratotype region, the Moscow Syneclise.[61]

The emphasis placed on carbonate successions of reduced thickness and containing stratigraphic breaks at which chronostratigraphic boundaries have been placed, creates problems of correlation with areas where sedimentation was continuous across these chronostratigraphic boundaries.[66] It is also apparent that stages and their subdivisions, the horizons, are largely linked to changes in the composition of shallow marine faunas, particularly fusulinid foraminifers. Where there are gaps in the record, either due to breaks in the sedimentation or as a result of facies changes, the exact ranges of key fossil taxa become poorly known and boundary problems will arise. It might have been better if the USSR stratotypes were designated in areas with more varied facies and more continuous sedimentation, e.g. the Donets Basin. It is also noted that the boundaries of major chronostratigraphic units, such as the Bashkirian, are changed to fit a biostratigraphic concept or to suit certain ideas on international correlation. The lack of stability in the contents of a chronostratigraphic unit as typified by a stated reference section (stratotype) tends to confuse later workers and will generally impair the usefulness of the stratigraphic unit.

4. North America

The essential characteristics of the North American classification is the division of the Carboniferous into two parts, the Mississippian and Pennsylvanian. These are generally regarded as systems. There is a widespread discontinuity between Mississippian and Pennsylvanian, and there are only a few localities where a transition occurs.[67-69] It is clear that this discontinuity is of such widespread regional extent that a large scale regression is involved, similar to that registered on the Russian Platform, another major stable area of continental dimensions. Pennsylvanian sedimentation commenced at different times in different parts of the area, and is partly more terrestrial in character.

The Mississippian is generally developed in marine facies and characteristically as carbonates. The type section is a composite of several localities in the Mississippi Valley. Four provincial series are recognised, viz. the Kinderhookian, Osagean, Meramecian, and Chesterian (Fig. 3). Correlations in the type area are based primarily on foraminifers (Mamet's zones 7-19),[71] but conodonts, brachiopods and corals have also played impor-

		Mid-West Continent	Foraminiferal Zones	Conodont Zones
PENNSYLVANIAN	Virgilian	Wabaunsee Shawnee Douglas		
	Missourian	Lansing Kansas City Pleasanton	Waeringella Kansanella Eowaerinella	
	Desmoinesian	Marmaton Cherokee	Beedeina Wedekindellina 23	Streptognathodus-dominated Neognathodus-dominated
	Atokan		22 Fusulinella 21 Parafusulinella	N. medadultimus N. bothrops Dipl. orphanus- D. coloradoensis Id. ouachitensis
	Morrowan	Winslow Bloyd Hale	20 Pseudostaffella Eostaffella	Id. convexus Id. klapperi I. delicatus N. bassleri N. symmetricus I. sinuatus Rhachistognathus primus
MISSISSIPPIAN	Chesterian		18/19 Archaediscus tenuis Millerella 17 16 Asteroarchaediscus warnanti	R. muricatus A. unicornis C. naviculus
	Meramecian	Ste Genevieve Lst. St Louis Lst Salem Shales Warsaw Shales	14/15 Zelleria 13 A. karreri 10-12 Nodosarchaediscus	G. bilineatus
	Osagean	Keokuk Lst. Burlington Lst. Fern Glen Lst. Meppen Lst.	8/9 Tetrataxis 7 Globoendothyra	G. texanus Sc. anchoralis-latus G. typicus
	Kinderhookian	Chouteau Lst. Hannibal Shale Glen Park Lst.	6	Si. isosticha-u. crenulata Si. cooperi Si. quadruplicata Si. crenulata Si. duplicata Siphonodella sulcata
U.D.		Louisiana Lst.		S. praesulcata

		Ammonoid Zones	Miospore Zones	Pennsylvanian Stratotype
PENNSYLVANIAN	Virgilian		Thymospora thiessenii Apiculatisporites lappites - Latosporites minutus	Monon-gahela
	Missourian		Punctatosporites minutus - P. obliquus	Conemaugh
	Desmoinesian	Gonioloboceras-Wellerites	Lycospora granulata- Cappasporites distortus Schopfites colchesterensis- Thymospora pseudothiesseni Cadiospora magna- Mooreisporites inusitatus	Charles-ton
	Atokan	Boesites Paralegoceras-Diaboloceras	Radiizonates difformis Torispora securis- Vestispora fenestrata Microreticulatisporites nobilis- Endosporites globiformis	Kanawha
	Morrowan	Axinolobus modulus B. branneri Vern. pygmaeus Ark. relictus Quin. henbesti Retites semiretia	Schulzospora rara- Laevigatosporites desmoinensis Lycospora pellucida	New River Poca-hontas
MISSISSIPPIAN	Ches-terian	Eumorphoceras bisulcatum Tumalites varians Goniatites granosus		
	Mera-mecian			
	Osagean	G. multiliratus G. americanus		
	Kinder-hookian	Protocanites lyoni		
U. D.				

Fig. 3 Chronostratigraphic classifications and biozonationp for North America [67-78]

tant roles.[70] The stratigraphically very significant goniatites[76] play a predominant role in the areas of condensed sedimentation, such as the Ozark Uplift. Floras have played only a minor part in Mississippian stratigraphy.

The base of the Pennsylvanian has been the subject of several recent papers. It is clear that Pennsylvanian sedimentation commenced at different times in different parts of the American Continent. A comparison between the goniatite zonations in the Ozark Platform region of Arkansas/Oklahoma and in Western Europe has shown up the large number of breaks in the platform succession.[76] This is the area of the Morrowan and Atokan chronostratigraphic units (with regard to the Atokan stratotype and concept see Sutherland and Manger[77]). Purely marine sedimentation is characteristic of the Mid-Continent and Rocky Mountain areas, with the exception of the Illinois and Michigan basins which are mixed marine and terrestrial. The Appalachian, Naragansett and Maritime Canadian Basins are predominantly to entirely terrestrial. Where marine facies occur throughout, the Morrowan, Atokan, Desmoinesian, Missourian, and Virgilian Series are used to subdivide the Pennsylvanian System. In the Appalachian Basin the Pottsville, Allegheny, Conemaugh, and Monongahela divisions have been used, but the project of the Pennsylvanian stratotype[79] has distinguished Lower, Middle and Upper Pennsylvanian Series subdivided into formational units as follows: Pocahontas, New River (Lower Pennsylvanian); Kanawha, Charleston (Middle); Conemaugh, Monongahela, Dunkard (in part) (Upper Pennsylvanian).

There is a presumption that Pennsylvanian strata follow conformably on Mississippian in the composite stratotype of the Appalachian Basin, and that the full Pennsylvanian System is represented.[79] Type sections have been described from West Virginia.[79] Plant megafossils are used for international correlation,[75] suggesting that the Pocahontas Formation is mainly middle Namurian, whilst New River would range from upper Namurian to basal Westphalian B. The presence of *Lyginopteris hoeninghausii* (Brongniart) Goepp. in both formations would suggest Westaphalian A, but the total flora is too poor to effect precise correlations. The Kanawha Formation commences in Westphalian B and possibly reaches into highest Westphalian C. The overlying Charleston Sandstone contains a Westphalian D flora. Problems develop with the Conemaugh Formation which is supposed to have a gradational contact with the Kanawha,[75] but which contains a floral assemblage suggesting a level as high as Stephanian B. A similar age may be assigned to the Monogahela. It is noted that the megafloras of the so-called Pennsylvanian stratotype have not yet been studied in detail, thus making it difficult to fit the records in the better known West European plant zonation. The presence of *Callipteris conferta* (Sternberg) Brongniart in the overlying Dunkard has been taken to mean Permian strata, but this fossil ranges upwards from Stephanian C in Western Europe. The apparent absence of lower Stephanian floras in the so-called Pennsylvanian stratotype is puzzling and raises the question of a possible gap in the succession.

Miospore assemblage zones have been used to correlate the Pennsylvanian section of the Illinois Basin (1 040m thick) with West European chronostratigraphic units.[74] A full succession of Westphalian and Stephanian strata seems to be present. Further correlations are based on marine faunas in the Illinois Basin which apparently allow recognition of Morrowan, Atokan, Desmoinesian, Missourian, and Virgilian provincial series. Problems occur in the upper part of the section where the stratigraphic gap between Westphalian D and Stephanian A in Saar-Lorraine has not been taken into account for the correlation, and the Westphalian/Stephanian boundary consequently placed too high in the Illinois succession. Megafloral remains show this boundary to lie within the Carbondale Formation,[20] i.e. within the Desmoinesian, and not within the lower Missourian, as Peppers[74] has suggested.

Originally brachiopod faunas were used to subdivide the marine cyclical deposits[80-82] of the Mid-Continent area, but fusulinid foraminifers and conodonts have become increasingly important. Differences in fusulinid nomenclature between workers in the United States and in the USSR have confused the issue but correlations with the USSR are now reasonably well established.[78] Several problems still remain, however. The presence of *Profusulinella* in the Atokan forms a major problem, but the cited occurrences may actually be morrowan in age.[83] Conodonts have not been used to their full potential since their distribution in the Pennsylvanian and especially in the Middle and Upper Carboniferous of the USSR is still incompletely known, but studies are in progress.

The stratigraphic information on the Mississippian and Pennsylvanian of the United States was summarised on the occasion of the Carboniferous Congress in Champaign-Urbana, in 1979.[84] A general review of North American correlations has been given by the COSUNA Project, which (for the Carboniferous) also provided approximate correlations with Western Europe and the USSR.[85]

5. China

Five major depositional areas are distinguished, viz. (1) South China, (2) North China, (3) Northwest China, (4) Xizang (Tibet)-West Yunnan, (5) Tianshan-Hinggan.[86]

The standard succession of chronostratigraphic units (Fig. 4) has been obtained from the South China region, an area bounded to the north by the Sino-Korean Platform and westwards by Xizang (Tibet)-West Yunnan. It slopes eastwards into the shelf sea bordering the Pacific Ocean. Marine carbonates are found with occasional terrestrial intercalations of local occurrence. Palaeotethyan invertebrate faunas allow correlations with the USSR above all. The Lower/Middle Carboniferous boundary is drawn at the base of the *Reticuloceras* Zone, as in the USSR. The Lower Carboniferous is subdivided into three stages, the Aikuanian, Tatangian and Dewunian, which correspond to the Tournaisian, Visean and Serpukhovian. Middle Carboniferous consists of the Huashibanian and Dalanian Stages. Although these stages collectively correspond to Bashkirian and Moscovian, their boundary is taken at a lower level, i.e. at the incoming of *Profusulinella*. The Middle/Upper Carboniferous boundary has been drawn at the incoming of *Triticites*, but since the preceding faunas are clearly late Moscoviasn in age a gap may be suspected, corresponding to the basal Kasimovian. The Upper Carboniferous Maping Formation (s.l.) reaches well into the Permian.

In the North China region Carboniferous sedimentation commenced in the Middle Carboniferous at a level corresponding to the Moscovian or Westphalian. Invertebrate

		Fusulinid Zones	Ammonoid Zones
L. PERMIAN	Mapingian	Pseudoschwagerina robusta- Zellia chengkungensis P. parabeedei - Sphaero- schwagerina sphaerica P. morsei - Robusto- schwagerina xiaodushanica	
UPPER CARB.	Xiaodushanian	T. sikhanensis compactus Triticites dictyophorus T. schwageriniformis Eotriticites montiparus Protriticites subschwagerinoides	
MIDDLE CARBONIFEROUS	Dalanian	Fusulina quasicylindrica F. cylindrica F. schellwieni Fusulinella praebocki Profusulinella	Gastrioceras aff. coramatum Branneroceras Eoparalegoceras Pseudoparalegoceras Neodimorphoceras
	Huashibanian	Pseudostaffella antiqua	Gastrioceras Branneroceras Reticuloceras Retites
LOWER CARB. (FENGNINIAN)	Alkuanian Tatangian Dewunian	Eostaffella hoshenensis	Homoceras cf. subglobosum Proshumardites uralicus Cravenoceras dewuense Gattendorfia Eocanites
U. DEV.	Shaod.		

Fig. 4 Chronostratigraphic classsification and biozonations for China[19,89]

faunas are characteristic for the Palaeotethys and the floras are Amerosinian and comparable with European assemblages. The Upper Carboniferous Taiyuan Formation shows Palaeotethys faunas and Amerosinian floras with the introduction of (Permian) Cathaysian elements. It is possible that the higher part of the Taiyuan Formation should be attributed to the Permian, there being a gradual transition without climatic changes.

Northwest China shows marine Lower Carboniferous deposits followed by mixed marine and terrestrial strata, again with Palaeotethyan faunas and Amerosinian plant fossils. The Upper Carboniferous contains faunas and floras similar to those obtained from the Taiyuan Formation.

Xizang (Tibet)-West Yunnan is still incompletely known, but marine Lower and Middle Carboniferous with Palaeotethyan faunas have been recorded. Glaciomarine diamictites have been found in the Jilong Formation on the northern slope of the Himalayas. Overlying marine deposits carry a Palaeotethyan fauna, but subsequent Permian strata contain *Glossopteris,* the Gondwana index. The Himalaya subregion belongs to the northern edge of the Gondwana Realm (and Continent) and is different from the rest of China which is Palaeotethyan and Amerosinian.

Tianshan-Hinggan is an area of mobile basins north of the stable platform areas with epicratonic basins which occupy most of China. Mainly siliciclastic and volcanoclastic rocks were laid down in Early and Middle Carboniferous times. Palaeotethyan faunas occur in these deposits which include some Moscovian limestones. Floral remains show links with the Angara Realm. Upper Carboniferous deposits are also quite thick and contain Palaeotethyan faunas and Angara plants. There is a gradual transition with the Permian.

The Palaeotethyan faunas found in most of the Chinese Carboniferous and the presence of carbonate platform facies over wide areas, particularly in South China, make the USSR classificatory scheme immediately applicable. Plant fossils of the Amerosinian Realm allow comparison with floras of the Caucasus, Donets Basin and throughout Central and Western Europe. Towards the end of the Carboniferous and throughout most of the Permian the Chinese floras continue to show evidence of a humid and warm climate which continued conditions that were widespread all over the palaeoequatorial belt in Carboniferous times. In most of Europe and North America conditions became more arid in Permian times, as a result of which generally poorer floral assemblages of a different complexion were found. These Euramerican (Atlantic) type floras are therefore not immediately comparable to the Chinese (Cathaysian) floral associations. A narrow strip along the northern slope of the Himalayas has been sutured onto the Asian Continent and belongs to the Gondwana Realm (and Continent). In the Tianshan-Hinggan region, in the north-northwest of China, a transition to the Angara Realm has been recorded.

6. Gondwana

Well-dated Carboniferous deposits laid down at high latitudes on the Gondwana Supercontinent are rare. Certain areas in the southern Himalayas qualify in this respect,[87] but the best known successions are in Australia and Argentina.[88-89] In Australia the Lower Carboniferous has been identified in West European chronostratigraphic terms on the basis of ammonoid faunas, conodonts, palynomorphs, etc. Attempts have been made to identify also Namurian, Westphalian and Stephanian, but these attempts cannot be regarded as successful. Independent zonations have been set up for the microfloras and other fossil groups, most notably the brachiopods. In Argentina,[90] the Paganzo Basin provides the best example of Carboniferous strata which may include the Lower Carboniferous and which certainly represent the Middle to Upper Carboniferous, leading into the Permian. The eastern part of the basin is epicratonic, with progressive onlap onto the Pampean Arch on its eastern border, whereas the western part probably grades into the largely marine successions of the Andean region (Precordillera and, above all, Cordillera Frontal). Plant fossils of Pennsylvanian age show the lack of diversity to be associated with high latitudes, and the absence of large trees (dwarf lycophytes have been recorded) ties in with the assumption that a tundra landscape may have been present. Evidence of glacial conditions includes the presence of dropstones in lacustrine deposits which alternate with the plant-bearing strata with coals. Marine invertebrate faunas include brachiopods, bivalves, gastropods, etc., but typically lack fusulinids and conodonts. These faunas are poorly diversified, and not immediately comparable with the Tethyan faunas of the palaeoequatorial belt. They have allowed a broad biozonation which is to be compared with that erected in Australia. Indeed, the biostratigraphic control is so coarse that palaeomagnetic reversal stratigraphy is used in conjunction with radiometric dating of key points in the succession in order to provide a stratigraphic framework that cannot possibly be as detailed as that in the palaeoequatorial belt.

7. Angara

This area, comprising Siberia, NE Kazakhstan and Mongolia, was a stable platform with epicratonic basins with mainly terrestrial sediments, particularly with regard to Middle and Upper Carboniferous. Marine basins existed on the margins of Angaraland. The area corresponded to a higher palaeolatitude than the broad palaeoequatorial belt of Carboniferous times. Similar to the Gondwana high latitude area, one finds that the Lower Carboniferous (with more common marine facies) can be identified readily in terms of palaeoequatorial belt chronostratigraphy. Likewise, the Middle and Upper Carboniferous show different floras and faunas which do not permit a ready identification with palaeoequatorial belt units. This can only be due to increased climatic differentiation according to palaeolatitude, and may be related to the Pennsylvanian Ice Age.

Correlations in marine strata are based on foraminiferal and brachiopod faunas, whereas the terrestrial deposits are zoned on plant fossils. Both faunas and floras show a progressive increase in endemic taxa upwards in the Middle and Upper Carboniferous succession, and it has been noted that floral endemism commenced earlier than faunal endemism. This is readily understandable since climatic changes will have a more immediate effect on biota living on the land surface. Four successive lepidophytalean plant assemblages have been described for the Lower Carboniferous of Angaraland. These are replaced rather suddenly by a pteridosperm assemblage; a floral change that is attributed to the so-called Ostrogian Cooling Episode[22] and which apparently corresponds to the major floral change in the higher part of the lower Namurian. This presumably marks the onset of the Pennsylvanian Ice Age and may be correlated with the widespread withdrawal of the sea from carbonate platform areas at the end of the Mississippian. Higher up in the succession the pteridosperm assemblage is replaced by cordaitean-pteridosperm and almost pure cordaitean assemblages which have often been regarded as having constituted a cold-temperate taiga forest cover.

With regard to chronostratigraphic units, the same constraints apply here as for the Gondwana Realm, although perhaps to a lesser degree. The relatively poor diversification of faunas and floras makes detailed correlations more difficult and makes it necessary to use broader chronostratigraphic units than can be distinguished in the tropical to subtropical palaeoequatorial belt.

The data and considerations mentioned above are largely derived from the recent synthesis provided by Durante and Meyen.[91]

8. Radiometric datings

The radiometric datings for the Carboniferous System are still highly controversial.[92-94] A reasonable appoximation for the base of the Carboniferous seems to be 360±10Ma, whilst its top is c.290Ma. An age of 320±10Ma for the mid-Carboniferous boundary is rather speculative because of the lack of good stratigraphic control for the volcanic rocks used for the radiometric age identification. The figure of 300Ma for the base of the Stephanian seems less conjectural but is still only a rough approximation.

CONCLUSIONS AND RECOMMENDATIONS

1. The evidence for climatic differentiation at the end of Mississippian times, which is reflected in the changes in floral and faunal assemblages and their distribution patterns, suggests a major world-wide event that should be reflected in the chronostratigraphic classification. This event is the onset of the Pennsylvanian Ice Age which also produced widespread regression from the shallow carbonate platforms of Mississippian times. Bouroz et al.[2] suggested that the Mississippian/Pennsylvanian boundary should be regarded as a subsystemic boundary and this suggestion is maintained by the present authors, in view of the fact that the Carboniferous is historically regarded as an entity. Otherwise, the American habit of distinguishing these major units as separate systems should be recommended.

The IUGS Subcommission on Carboniferous Stratigraphy has recently focused on this mid-Carboniferous boundary,[21,24,27] providing a biostratigraphic definition of this boundary but staying its hand with regard to a boundary stratotype and definition of the first chronostratigraphic unit above the boundary (apart from the Pennsylvanian). From the biostratigraphic data presented, it appears that the floral change in the palaeoequatorial belt is slightly in advance of the faunal changes (late Arnsbergian as against Chokierian-Alportian), and this conforms to the pattern observed in the Angara Realm, in the northern hemisphere temperate belt.

2. During the Mississippian the climatic differentiation commenced but there is no abrupt change at the Devonian/Carboniferous boundary. The IUGS Devonian-

Carboniferous Boundary Working Group has recommended a biostratigraphic level (at the transition from *Siphonodella proesulcata* to *S. sulcata*) but has refrained, thus far, from recommending a boundary stratotype.[5] The recommended level is immediately below that recommended at the Second Heerlen Congress, which was at the incoming of *Gattendorfia*.[4]

3. Bouroz et al.[2] recommended a series boundary between Middle and Upper Carboniferous, and subdivision into Lower, Middle and Upper Carboniferous was recommended for provisional world-wide use by the IUGS Subcommission on Carboniferous Stratigraphy.[95] The exact position of this boundary has been left open, but it is noted that the Westphalian/Stephanian boundary in Western Europe lies below both the Desmoinesian/Missourian boundary in the United States and the Moscovian /Kasimovian boundary of the USSR. The Middle/Upper Carboniferous boundary is obviously less important than the Lower/Middle Carboniferous boundary, but it should be worthwhile to examine the various different groups of organisms for apparent changes near this boundary. It is noted that the main floral change occurs in mid-Westphalian D and the fusulinid biostratigraphy shows the introduction of *Protriticites* at a higher level (base of the Kasimovian). The Westphalian/Stephanian boundary in Western Europe has originally been drawn at unconformable contacts, and is presenly drawn in a continuous succession where no major biological change exists.

The Middle/Upper Carboniferous boundary cannot be distinguished unequivocally in the Gondwana and Angara Realms. In view of the difficulties of correlation across palaeolatitudinal belts in parts of the geological column showing a marked climatic differentiation, it may be that the Middle/Upper Carboniferous boundary cannot be recognised world-wide unless palaeomagnetic reversals are used as events allowing the stages (and this series boundary) to be fitted into a broad framework.

4. The Carboniferous/Permian boundary is taken at different places in different parts of the world. Like the base of the Pennsylvanian, the base of the Permian may coincide with a climatic event; this time the waning of the Ice Age. This seems the only way to determine a boundary that can be made to serve world-wide. In the eastern part of the European USSR, traditionally regarded as the Lower Permian type area, the base of the Asselian is the preferred location for the Carboniferous/Permian boundary.[12] This is correlated with the base of the West European Autunian as redefined by Bouroz and Doubinger[9] (originally the lower/middle Autunian boundary) on the basis of a climatic change as shown by the rather sudden increase in the proportion of *Potonieisporites*. This is mainly a conifer pollen and the increase in its relative abundance means the widespread elimination of swamp and wet floodplain habitats which characterised the Carboniferous of the palaeoequatorial belt. The same climatic change is picked up in the reverse order in the Gondwana Realm where the *Potonieisporites* Zone is followed by the *Cristatisporites* Zone[8,21] as a result of the amelioration of climatic conditions at the end of the Ice Age. Its local effect is different, but the same event seems to be the trigger, and this allows world-wide correlation.

INT.	USSR		WESTERN EUROPE		NORTH AMERICA		CHINA
L.P.	L.P. ASS.	Sokoljegorsky	Lower Permian Autunian		L.P.	Lower Permian	UPPER CARBONIFEROUS Maping
PENNSYLVANIAN UPPER CARBONIFEROUS	UPPER CARB. GZHELIAN	Noginsky	SILESIAN Stephanian	C	PENNSYLVANIAN Virgilian	Wabaunsee	
		Pavlovo-Posadsky				Shawnee	
		Amerevsky				Douglas	
		Rusavkinsky		B			
	KASIMOVIAN	Yauzsky			Missourian	Lansing	UPPER CARBONIFEROUS Xiaodushanian
		Dorogomilovsky				Kansas City	
		Khamovnichesky		(A) Barruelian		Pleasanton	
		Krevyakinsky					
				Cantabrian			
PENNSYLVANIAN MIDDLE CARBONIFEROUS	MIDDLE CARB. MOSCOVIAN	Myachkovsky	Westphalian	D	Desmoinesian	Marmaton	MIDDLE CARBONIFEROUS Dalonian
		Podolsky		(C) Bolsovian			
		Kashirsky		(B) Duckmantian	Atokan		
		Tsninsky					
		Vereisky					
	BASHKIRIAN	Melekessky		(A) Langsettian			
		Cheremshansky	Namurian	Yeadonian	Morrowan	Winslow	
		Prikamsky		Marsdenian		Bloyd	MIDDLE CARBONIFEROUS Huashi-banian
		Severokeltmensky		Kinderscoutian		Hale	
		Krasnopolyansky					
MISSISSIPPIAN LOWER CARBONIFEROUS	LOWER CARBONIFEROUS SERPUKHOV.	Voznesensky		Alportian	MISSISSIPPIAN Chesterian	Elvirian	Dewunian
		Zapaltyubinsky		Chokierian			
		Protvinsky		Arnsbergian			
		Steshevsky		Pendleian		Hombergian	
		Tarussky					
	VISEAN	Venevsky	DINANTIAN Viséan	Brigantian		Gasperian	LOWER CARB. (FENGNINIAN) Tatangian
		Mikhailovsky		Asbian			
		Aleksinsky			Meramecian	Ste Genevieve Lst.	
		Tulsky		Holkerian		St Louis Lst.	
						Salem Shales	
		Bobrikovsky		Arundian		Warsaw Shales	
		Radaevsky		Chadian			
		Elkhovsky			Osage	Keokuk Lst.	
	TOURNAISIAN	Kizelovsky	Tournaisian	Ivorian		Burlington Lst.	Aikuanian
						Fern Glen Lst.	
						Meppen Lst.	
		Cherépetsky		Hastarian		Chouteau Lst.	
		Upinsky				Hannibal Shale	
		Malevsky				Glen Park Lst.	
U.D.		Zavolzhsky					
	U.D. FAM.		U.D. Fam.	Strunian	U.D.	Louisiana Lst.	U.D. Shaod.

Fig. 5 Correlation chart for the major chronostratigraphic subdivisions of North America, Western Europe, USSR, and China (see Figs. 1-4)[96-97]

5. Stages within the three major units, lower, middle and upper, are presently two per series in the USSR and four to eight per series in Western Europe (see Fig. 5). The stage being the basic chronostratigraphic unit, it should be recognisable world-wide on its total fossil contents. Since the Middle and Upper Carboniferous show much more diversified floras and faunas in the palaeoequatorial belt than occur in the Gondwana and Angara Realms, the size of the constituent stages can be much smaller in the palaeoequatorial belt than in the two other areas mentioned.

Within the palaeoequatorial belt it should be possible to recognise three stages in the Lower Carboniferous, for which Tournaisian, Visean and Serpukhovian are recommended (in agreement with Bouroz et al.[2]). The Middle Carboniferous can be subdivided into four or five stages with a duration of approximately 5 to 6 million years each. The Bashkirian and Moscovian Stages of the USSR classification are obviously too large. Besides, the position of the Bashkirian/Moscovian boundary is subject to discussion. For the Upper Carboniferous two stages can be usefully distinguished, corresponding approximately to lower and upper Stephanian, i.e. Cantabrian + Barruelian and Stephanian B + C of Western Europe. The Kasimovian and Gzhelian of the USSR are rather unequal in size (see Fig. 5).

6. It is noted that a certain tendency to place stage boundaries at stratigraphic breaks (numerous examples exist) can only lead to problems of recognition of such boundaries where continuous sedimentation occurred across the boundary. The base of the Pennsylvanian in the USA, the base of the Moscovian in the USSR and the base of the Stephanian in NW Europe are examples of the problems that arise in such cases. The IUGS Subcommission on Stratigraphic Nomenclature has already recommended that chronostratigraphic boundaries should be drawn in continuous successions.[98] One may add that stable carbonate platform successions are notoriously incomplete. Where small breaks exist that cannot be detected palaeontologically, they may perhaps be ignored, but who knows whether the power of resolution of certain biozonations may not show up such small breaks at a later data? Therefore, it may be wise to avoid these carbonate platform successions for international stratotypes.

ACKNOWLEDGEMENTS

Dr. A.C. Higgins is thanked for details of the conodont zonations. The comments of Dr. B. Owens are also gratefully acknowledged.

REFERENCES

1. Lane HR and Manger WL, Cour Forch-Inst Senckenberg, 74 (1985), 15.
2. Bouroz A, Einor OL, Gordon M, Meyen SV and Wagner RH, Izv Akad Nauk SSSR, otd Geol, 1977, 2, 5; Industrie min, 60 (10) (1978), 469; 8ᵉ Congr Int Strat Geol Carb Moskva, 1975, C R 1 (1978), 36.
3. Boucot AJ, Developments Palaeont Strat, 1 (1975), 346.

4. Jongmans WJ and Gothan W, 2^{e} Congr Int Strat Geol Carb Heerlen, 1935, C R 1 (1937), 1.
5. Paproth E and Streel M (eds), Cour Forsch-Inst Senckenberg, 67 (1984).
6. Reitlinger EA, Semichatova SV, Byvsheva TV, Chizhova VA, Kononova LI, Lipina OA, Aisenverg DE, Antropov IA, Brazhnikova NE, Durkina AV, Fedorova TI, Grozdilova LP, Kedo GJ, Khalymbadzha VG, Lapina NN, Martynova NM, Nasikanova OH, Nenastieva VE, Nechaeva MA, Rozhdestvenskaya AA. Samoilova RB, Tkacheva UD and Umnova BT, 8^{e} Congr Int Strat Geol Carb Moskva, 1975, C R 1 (1978), 70;Yorkshire Geol Soc Occas Publ, 4 (1979), 23.
7. Rui Lin and Zhang Linxin, Jour Strat, 10 (4) (1986), 249.
8. Wagner RH, Actas II Congreso Argent Paleont Bioestrat y I Congreso Latinoamericano Paleont, Buenos Aires, 1978, IV (1980), 177.
9. Bouroz A and Doubinger J, In: Symposium on Carboniferous Stratigraphy, Geol Surv Prague (1977), 147.
10. Murchison RI, Verneuil E de and Keyserling A von, The geology of Russia in Europe and the Ural Mountains, I Geology, J Murray (P Bertrand) (1845).
11. Ruzhencev VE, Dokl Akad Nauk SSSR, 99, 6 (1954), 1079.
12. Rauser-Chernousova DM, Shchegolev AK, Bensh FR, Kireeva GD, Glushenko NV, Il'ina TG, Morozova IP, Meyen SV, Luber AA, Podoba BG, Faddeeva IZ and Lapkin IYu, Yorkshire Geol Soc Occas Publ, 4 (1979), 175.
13. Kozur H, Freib Forsch, C 319 (1977), 79.
14. Zhang Zuqi, Newsl Strat, 13 (3) (1984), 156.
15. Barkhatova VN, Trudy Mezhved Strat Komit SSSR, 4 (1970), 167.
16. Rauser-Chernoussova DM, 4^{e} Congr Int Strat Geol Carb Heerlen, 1958, C R 3 (1962), 577.
17. Bensh FR, Stratigrafiya i fuzulinidy verkhnego paleozoya Yuzhnoj Fergany, Izdat. FAN (1972).
18. Li Xingxue and Zhang Linxin, In: The Carboniferous of the World, I, IUGS Publ, 16 (1983), 115.
19. Zhou Tieming, Sheng Jinzhang and Wang yujing, Acta Micropalaeont Sinica, 4 (2) (1987), 123.
20. Wagner RH, 9^{e} Congr Int Strat Geol Carb Washington and Champaign-Urbana, 1979, C R 2 (1984), 109.
21. Archangelsky S and Marques Toigo M, Actas II Congreso Argent Paleont Bioestrat y I Congreso Latinoamericano Paleont, Buenos Aires, 1978, IV (1980), 221.
22. Meyen SV, In: Palaozoische und mesozoische Floren Eurasiens und die Phytographie dieser Zeit, G. Fischer Verlag (1978), 26.
23. Wagner RH, Saunders WB and Manger WL, 10^{e} Congr Int Strat Geol Carb Madrid, 1983, C R 1 (1985), 59.
24. Lane HR, Bouckaert J, Brenckle P, Einor OL, Havlena V, Higgins AC, Yang Jingzhi, Manger WL, Nassichuk W, Nemirovskaya T, Owens B, Ramsbottom WHC, Teitlinger EA and Weyant M, 10^{e} Congr Int Strat Geol Carb Madrid, 1983, C R 4 (1985), 323.
25. Havlena V, In: Biostratigraphic Data for A Mid-Carboniferous Boundary, Subcom Carboniferous Strat (1982), 112.
26. Wagner RH, In: Biostratigraphic Data for A Mid-Carboniferous Boundary, Subcom Carboniferous Strat (1982), 120.
27. Ramsbottom WHC, Saunders WB and Owens B (eds), Biostratigraphic Data for A Mid-Carboniferous Boundary, Subcom Carboniferous Strat (1982).
28. Wagner RH, Bull Soc belge Geol, 83 (1974) (1976), 235.
29. Olaussen S, Geol Mag, 118 (3) (1981), 281.
30. Bergstrom J, Bless MJM and Paproth E, Z dt geol Ges, 136 (1985), 181.
31. Teichmuller M, 4^{e} Congr Int Strat Geol Carb Heerlen, 1958, C R 3 (1962), 699.

32. Bless MJM and Winkler Prins CF, 7e Congr Int Strat Geol Carb Krefeld, 1971, C R 1 (1972), 231.
33. Calver MA, 6e Congr Int Strat Geol Carb Sheffield, 1967, C R 1 (1969), 233.
34. Leckwijck WP van, 5e Congr Int Strat Geol Carb Paris, 1963, C R 1 (1964), 37.
35. Ramsbottom WHC, Proc Yorkshire Geol Soc, 39, 4, 28 (1973), 567; 41, 3, 24 (1977), 261.
36. Ross CA and Ross JRP, Geology, 13 (1985), 194.
37. Laveine JP, In: Symposium on Carboniferous Stratigraphy, Geol Surv Prague (1977), 71.
38. Clayton G, Coquel R, Doubinger J, Gueinn KJ, Loboziak S, Owens B and Streel M, Meded Rijks Geol Dienst, N S, 29 (1977).
39. Clayton G, 10e Congr Int Strat Geol Carb Madrid, 1983, C R 4 (1985), 9.
40. Trueman AE and Weir J, Palaeontogr Soc Mon, (1946), 29.
41. Conil R, Longerstaey PJ and Ramsbottom WHC, Mem Inst Geol Univ Louvain, 30 (1979).
42. Ramsbottom WHC, Calver MA, Eagar RMC, Hodson F, Holliday DW, Stubblefield CJ and Wilson RB, Geol Soc Spec Rep, 10 (1978).
43. Owens B, Riley NJ and Calver MA, 10e Congr Int Strat Geol Carb Madrid, 1983, C R 4 (1985), 461.
44. Wagner RH and Winkler Prins CF, An Fac Cienc Porto, Suppl Vol 64 (1983) (1985), 359.
45. George TN and Wagner RH, 7e Congr Int Strat Geol Carb Krefeld, 1971, C R 1 (1972), 139.
46. Ramsbottom WHC (ed), Field Guide to the Boundary Stratotypes of the Carboniferous Stages in Britain, Subcom Carboniferous Strat (1981).
47. Anonymous, Field Guide to the Boundary Stratotypes of the Carboniferous Stages in Ireland, Subcom Carboniferous Strat (1981).
48. Riley NJ, Razzo MJ and Owens B, 10e Congr Int Strat Geol Carb Madrid, 1983, C R 1 (1985), 35.
49. Bertrand P, 2e Congr Int Strat Geol Carb Heerlen, 1935, C R 1 (1937), 67.
50. Jongmans WJ and Pruvost P, Bull Soc Geol France, 5 (20) (1950), 335.
51. Doubinger J, Mem Soc geol France, 35, N S, 75 (1956), 1.
52. Bouroz A, C R Acad Sci Paris, 266 (1968), 2219.
53. Bouroz A and Doubinger J, Industrie min, 60 (10)(1978), 485.
54. Kahler F, Carinthia II, 176 (1986), 1.
55. Wagner RH and Bowman MBJ, Newsl Strat,12 (3)(1983), 132.
56. Aisenverg DE, Brazhnikova NE, Vassilyuk NP, Vdovenko MV, Gorak SV, Dunaeva NN, Zernetskaya NV, Poletaev VI, Potievskaya PD, Rotai AP and Sergeeva MT, 8e Congr Int Strat Geol Carb Moskva, 1975, C R 1 (1978), 158; Yorkshire Geol Soc Occas Publ, 4 (1979), 197.
57. Fissunenko OP and Laveine JP, 9e Congr Int Strat Geol Carb Washington and Champaign-Urbana, 1979, C R 1 (1984), 95.
58. Reitlinger EA, Vdovenko MV, Gubareva VS and Shcherbakov OA, In: The Carboniferous of the World III, IUGS Publ (in prep).
59. Einor OL, In: The Carboniferous of the World III, IUGS Publ (in prep).
60. Solovieva MN, Gubareva VS, Ivanova EA, Fissunenko OP, Shcherbakov OA and Einor OL, In: The Carboniferous of the World III, IUGS Publ (in prep).
61. Grigorieva AD, Shchegolev AK, Alexeeva IA and Shcherbakov MV, In: The Carboniferous of the World III, IUGS Publ (in prep).
62. Solovieva MN, Fisunenko OP, Goreva NV, Barskov IS, Gubareva VS, Dzhenchuraeva AV, Dalmatskaya II, Ivanova EA, Poletaev VI, Popov AV. Rumyantseva ZS, Teteryuk VK and Shik EM, 10e Congr Int Strat Geol Carb

Madrid, 1983, C R 1 (1985), 11.
63. Solovieva MN, 10e Congr Int Strat Geol Carb Madrid, 1983, C R 1 (1985), 21.
64. Pegusheva LM, Trudy VNIGNI, 52 (1968), 48.
65. Granados LF, Solovieva MN, Reitlinger EA and Martinez-Diaz C, 10e Congr Int Strat Geol Carb Madrid, 1983, C R 1 (1985), 27.
66. Solovieva MN, Fisunenko OP, Goreva NV, Barskov IS, Gubareva VS, Dzhenchuraeva AV, Dalmatskaya II, Ivanova EA, Poletaev VI, Popov AV, Rumyantseva ZS, Teteryuk VK and Shik EM, 10e Congr Int Strat Geol Carb Madrid, 1983, C R 1 (1985), 16.
67. Skipp B, Baesemann JF and Brenckle PL, 10e Congr Int Strat Geol Carb Madrid, 1983, C R 4 (1985), 403.
68. Lane HR, Baesemann JF, Brenckle PL and West RR, 10e Congr Int Strat Geol Carb Madrid, 1983, C R 4 (1985), 429.
69. Gordon M, Henry TW and Mamet BL, 10e Congr Int Strat Geol Carb Madrid, 1983, C R 4 (1985), 441.
70. Dutro JT, Gordon M and Huddle JW, US Geol Surv Prof Paper, 1010 (1979), S407.
71. Lane HR and Baesemann JF, In: Biostratigraphic Data for A Mid-Carboniferous Boundary, Subcom Carboniferous Strat (1982), 6.
72. Mamet BL and Skipp B, 6e Congr Int Strat Geol Carb Sheffield, 1967, C R 3 (1970), 1129.
73. Mamet BL, Skipp B and Sando WJ, Bull Amer Assoc Petrol Geol, 55(1) (1971), 20.
74. Peppers RA. 9e Congr Int Strat Geol Carb Washington and Champaign-Urbana, 1979, C R 2 (1984), 438.
75. Gillespie WH and Pfefferkorn HW, In: 9e Congr Int Strat Geol Carb Washington and Champaign-Urbana, 1979, Guidebook Field-trip 1 (1979), 86.
76. Gordon M, Jr, US Geol Surv Prof Paper, 460 (1964), 1.
77. Sutherland PK and Manger WL (eds), Oklahoma Geol Surv Bull, 136 (1984), 1.
78. Wilde GL, 9e Congr Int Strat Geol Carb Washington and Champaign-Urbana, 1979, C R 2 (1984), 543.
79. Englund KJ, Arndt HA, Schweinfurth SP and Gillespic WH, In: Geol Soc America Centennisal Field Guide, Southeastern Section (1986), 59.
80. Merriam DF (ed), Bull State Geol Surv Kansas, 169 (1964).
81. Wright CR, US Geol Surv Prof Paper, 853 (1975), 73.
82. Heckel PH, 9e Congr Int Strat Geol Carb Washington and Champaign-Urbana, 1979, C R 3 (1984), 535.
83. Shaver RH, In: Oklahoma Geol Surv Bull, 136 (1984), 101.
84. US Geol Surv Prof Paper, 1110 (1979), A-L; M-DD.
85. Salvador A, Amer Assoc Petrol Geol Bull, 69 (1) (1985), 181.
86. Yang Shihpu, Gao Lianda and Li Xingxue, In: The Carboniferous of the World, I, IUGS Publ, 16 (1983), 11.
87. Gupta VJ, Waterhouse JB and Bhargava ON, In: The Carboniferous of the World, II, IUGS Publ, 20 (1983), 147.
88. Roberts J, In: The Carboniferous of the World, II, IUGS Publ, 20 (1983), 9.
89. Rocha Campos AC and Archangelsky S, In: The Carboniferous of the World, II IUGS Publ, 20 (1983), 175.
90. Archangelsky S (ed), El Sistema Carbonifero en la Republica Argentina (Sintesis), Subcom Carboniferous Strat, Cordoba (1986).
91. Durante MV and Meyen SV, In: The Carboniferous of the World, III, IUGS Publ (in prep).
92. Winkler Prins CF, Newsl Carb Strat, 3 (1982), 14.
93. Forster SC and Warrington G, Mem Geol Soc London, (1981), 99.
94. Odin GS, Mem Geol Soc London, (1981), 114.
95. Bouroz A, Wagner RH and Winkler Prins CF, 8e Congr Int Strat Geol Carb Moskva, 1975, C R 1 (1978), 30.

96. Winkler Prins CF, Newsl Carb Strat, 2 (1981), 11.
97. Ross CA, In: Treatise on Invertebrate Paleontology, Geol Soc America and Univ Kansas Press (1979), 256.
98. Hedberg HD (ed), International Stratigraphic Guide, J Wiley and Sons (1976).

ACKNOWLEDGEMENTS

The financial contributions and administrative supports from the following organizations are gratefully acknowledged:

Co-sponsors

THE CHINA ASSOCIATION FOR SCIENCE AND TECHNOLOGY
THE PALAEONTOLOGICAL SOCIETY OF CHINA
THE GEOLOGICAL SOCIETY OF CHINA
THE CHINESE SOCIETY OF COAL
THE CHINESE SOCIETY OF PETROLEUM

Contributors

THE NATIONAL NATURAL SCIENCE FOUNDATION OF CHINA
THE ASSOCIATION OF GEOLOGISTS FOR INTERNATIONAL DEVELOPMENT
THE INTERNATIONAL UNION OF GEOLOGICAL SCIENCES
THE THIRD WORLD ACADEMY OF SCIENCES

Non-profit co-sponsors

THE PERMANENT INTERNATIONAL COMMITTEE FOR THE ICC
THE ALL-CHINA COMMISSION OF STRATIGRAPHY